NEUROANATOMY RESEARCH AT THE LEADING EDGE

CRANIAL NERVES

ANATOMY, FUNCTION AND CLINICAL SIGNIFICANCE

NEUROANATOMY RESEARCH AT THE LEADING EDGE

Additional books and e-books in this series can be found on Nova's website under the Series tab.

NEUROANATOMY RESEARCH AT THE LEADING EDGE

CRANIAL NERVES

ANATOMY, FUNCTION AND CLINICAL SIGNIFICANCE

THOMAS M. YI
EDITOR

Copyright © 2021 by Nova Science Publishers, Inc.

All rights reserved. No part of this book may be reproduced, stored in a retrieval system or transmitted in any form or by any means: electronic, electrostatic, magnetic, tape, mechanical photocopying, recording or otherwise without the written permission of the Publisher.

We have partnered with Copyright Clearance Center to make it easy for you to obtain permissions to reuse content from this publication. Simply navigate to this publication's page on Nova's website and locate the "Get Permission" button below the title description. This button is linked directly to the title's permission page on copyright.com. Alternatively, you can visit copyright.com and search by title, ISBN, or ISSN.

For further questions about using the service on copyright.com, please contact:
Copyright Clearance Center
Phone: +1-(978) 750-8400   Fax: +1-(978) 750-4470   E-mail: info@copyright.com

NOTICE TO THE READER

The Publisher has taken reasonable care in the preparation of this book, but makes no expressed or implied warranty of any kind and assumes no responsibility for any errors or omissions. No liability is assumed for incidental or consequential damages in connection with or arising out of information contained in this book. The Publisher shall not be liable for any special, consequential, or exemplary damages resulting, in whole or in part, from the readers' use of, or reliance upon, this material. Any parts of this book based on government reports are so indicated and copyright is claimed for those parts to the extent applicable to compilations of such works.

Independent verification should be sought for any data, advice or recommendations contained in this book. In addition, no responsibility is assumed by the Publisher for any injury and/or damage to persons or property arising from any methods, products, instructions, ideas or otherwise contained in this publication.

This publication is designed to provide accurate and authoritative information with regard to the subject matter covered herein. It is sold with the clear understanding that the Publisher is not engaged in rendering legal or any other professional services. If legal or any other expert assistance is required, the services of a competent person should be sought. FROM A DECLARATION OF PARTICIPANTS JOINTLY ADOPTED BY A COMMITTEE OF THE AMERICAN BAR ASSOCIATION AND A COMMITTEE OF PUBLISHERS.

Additional color graphics may be available in the e-book version of this book.

Library of Congress Cataloging-in-Publication Data

ISBN: 978-1-53618-823-3

Published by Nova Science Publishers, Inc. † New York

CONTENTS

Preface vii

Chapter 1 Clinical Anatomy of the Olfactory
Nerve, Bulb and Tract 1
Serghei Covanțev, Olga Belic and Ilia Catereniuc

Chapter 2 Cranial Nerve II: Optic Nerve; Anatomy,
Evaluation, Pathology and Surgical Approaches 41
Natalie Homer, Aliza Epstein and Craig Kemper

Chapter 3 Cranial Nerve VII: Anatomy,
Function and Clinical Significance 73
Akhil Surapaneni and Craig Kemper

Chapter 4 Vagus Nerve: The Clinical Importance
in the Metabolic Disorders 117
*Berrin Zuhal Altunkaynak,
Işınsu Alkan and Cengiz Baycu*

Index 157

PREFACE

Cranial Nerves: Anatomy, Function and Clinical Significance opens with a summary of the current data on the clinical anatomy and developmental anomalies of the first cranial nerve, the olfactory nerve.

Following this, the authors provide an overview of the second cranial nerve, the optic nerve, which is a vital component of the visual pathway.

The seventh cranial nerve, the facial nerve, which contains the somatic motor and visceral motor, as well as special sensory and general sensory fibers is discussed.

The 10th cranial nerve, the vagus nerve, is explored in closing, focusing on its motor functions responsible for the innervations of the outer ear canal, pharynx, larynx, heart, lung, gastrointestinal tract, stomach, pancreas and liver.

Chapter 1 - Examination of the cranial nerves is an important part of complete neurological assessment of the patient. With the development of imaging techniques, there is an increased awareness of the possible anatomical variations and developmental anomalies of the nervous system, which play a major role in neurology and neurosurgery. Although there is extensive data on the anatomical variations of some of the cranial nerves, olfactory nerve is a relatively understudied subject. There are several types of developmental anomalies that are encountered in clinical practice such as aplasia, hypoplasia, olfactory bulb ventricles, and duplication/

triplication of the olfactory bulb. While some of them present only structural defects, other play an important role in several diseases or serve as marker for further evaluation of the patient. Moreover, there are several genetic conditions, in which olfactory system malformation are common and require extensive workup. The current data about the olfactory nerve, bulb and tract are important in neurosurgery and neurology since these structures are often implicated in the diseases of the nervous system. Nevertheless, taking into account the advances in the understanding of the olfactory system and its relationship with other systems, the list of implicated specialists includes otolaryngologists, craniofacial surgeons, respiratory medicine physicians, endocrinologists, geneticists and psychiatrists. Therefore, the purpose of this chapter is to sum up the current data on the clinical anatomy and developmental anomalies of olfactory nerve, bulb, tract as well as their link to other conditions.

Chapter 2 - The second cranial nerve is the optic nerve and is a vital component of the visual pathway. The optic nerve spans from the eye to the optic chiasm where the two optic nerves fuse. The optic nerve can be divided into its various segments including the intraocular, intraorbital, intracanalicular, and intracranial portions of the nerve. Each segment has a unique blood supply based on location. The structure and function of the optic nerve can be analyzed by a number of means. The optic nerve head can be examined directly by ophthalmoscopic exam or can be visualized by a variety of imaging modalities including optical coherence tomography (OCT), confocal scanning laser ophthalmoscopy (CSLO), and scanning laser polarimetry (SLP). Neuro-imaging with computed tomography (CT) and magnetic resonance imaging (MRI) can also provide visualization of the orbit and optic nerve. A variety of tests can provide information on the functioning of the optic nerve. These include visual acuity testing, the pupillary exam, color vision assessment, and visual field testing.

Chapter 3 - The seventh cranial nerve (CNVII), the facial nerve, contains somatic motor, visceral motor, special sensory, and general sensory fibers. CNVII controls the muscles of facial expression along with the posterior belly of the digastric, stylohyoid, and stapedius. It gives parasympathetic innervation to several salivary glands and the nasal mucosa and provides taste sensation from the anterior two-third of the tongue. The superior portion of the facial nucleus receives bihemispheric innervation whereas the inferior portion only receives contralateral input. Second order neurons from the facial nucleus join and represent the main bulk of the facial nerve. The nervus intermedius, a branch of the facial nerve, carries general and special sensory fibers. The facial nerve exits at the superior olivary sulcus and joins a nerve bundle made up of the superior and inferior vestibular nerves along with the auditory nerve before entering the internal auditory meatus of the facial canal in the petrosal portion of the temporal bone. The course through the temporal bone is a complex route and predisposes CNVII to various pathologies unique to this anatomy. Facial nerve palsy can occur due to lesions along its entire anatomy. Traumatic etiologies include injuries during delivery, surgical trauma, penetrating parotid injuries, facial and temporal bone fractures. Numerous inflammatory and infectious etiologies contribute to facial nerve palsies. While primary tumors of the facial nerve are rare, neoplastic compression by vestibular schwannomas and other cerebellopontine angle tumors is more common. Perineural spread of malignant tumors can occur anywhere along the course of CNVII. Metabolic, inflammatory and infectious disorders are also considered in the pathology of CNVII. Clinical evaluation of a patient with a facial palsy begins with a history and physical examination. Timeline, disease progression, associated symptoms, and past medical history guide the examiner towards a diagnosis. The examiner should pay close attention to the external auditory canal and the surrounding structures. The House-Brackmann scale can be used to evaluate the degree of facial nerve paralysis. For patients with total paralysis of the facial nerve, electromyography and electroneurography are used to quantitate innervation to facial muscle and its response to stimulation. High resolution MRI and CT scanning can diagnose and evaluate facial nerve disorders as well as guide perioperative planning and assess the resolution of the lesion. Following traumatic injury to the facial nerve, surgery or medical therapy alone can improve facial nerve function. In cases of nerve impingement by surrounding structures, decompression can be performed whereas interpositional nerve grafting is used for transecting nerve injuries. Oculoplastic approaches improve and maintain ocular function in facial nerve dysfunction. Facial plastic approaches are frequently used in reanimation procedures to improve function and cosmesis.

Chapter 4 - The vagus nerve is the 10th cranial nerve with motor functions responsible for the innervations of the outer ear canal, pharynx, larynx, heart, lung, gastrointestinal tract, stomach, pancreas and liver. The vagus nerve can be evaluated as a regulator of body metabolism by receiving signals from the brain. Signals from the hypothalamus, the master chief of the body, are transmitted by the vagus to most of the peripheral organs. The bridge function between the brain and peripheral organs causes the vagus to be used as a treatment tool in metabolic disorders. The treatments are based on the vagus nerve as stimulation and blockade can be used in obesity, neural and metabolic diseases and diabetes. While the vagal stimulation is used in diseases such as obesity and blood sugar regulation; vagal blockade is used in the treatment of obesity, metabolic and neuronal diseases. The speed of the effect of vagal stimulation on each organ is very slow, especially in cases where biochemical reactions occur. For fast action, vagal blockage can be preferred. In addition, vagal effects on the pancreas in the gastrointestinal tract could initiate the inflammatory pathways. Although it does not receive direct innervation by the vagus; these pathways can induce same changes in the spleen. When planning vagal treatments, it is necessary to consider whether the treated organ which is innervated with the left or right branch of the vagus nerve. Another important point of the planning is which regions of these organs are innervated by which branch. For example, a planned study to evaluate the effects of the left cervical vagus on the lung will not yield any results. Therefore, vagus nerve should be stimulated from the correct segment and branch in order to perform an effective treatment. In this section, the effects of vagal stimulation and blockade on different organs and their effects on metabolic diseases will be discussed.

In: Cranial Nerves
Editor: Thomas M. Yi

ISBN: 978-1-53618-823-3
© 2021 Nova Science Publishers, Inc.

Chapter 1

CLINICAL ANATOMY OF THE OLFACTORY NERVE, BULB AND TRACT

Serghei Covanțev[*], Olga Belic and Ilia Catereniuc

Department of Human Anatomy, State University of Medicine and Pharmacy "Nicolae Testemițanu", Chișinau, Republic of Moldova

ABSTRACT

Examination of the cranial nerves is an important part of complete neurological assessment of the patient. With the development of imaging techniques, there is an increased awareness of the possible anatomical variations and developmental anomalies of the nervous system, which play a major role in neurology and neurosurgery. Although there is extensive data on the anatomical variations of some of the cranial nerves, olfactory nerve is a relatively understudied subject. There are several types of developmental anomalies that are encountered in clinical practice such as aplasia, hypoplasia, olfactory bulb ventricles, and duplication/triplication of the olfactory bulb. While some of them present only structural defects, other play an important role in several diseases or

---

[*] Corresponding Author's E-mail:kovantsev.s.d@gmail.com.

serve as marker for further evaluation of the patient. Moreover, there are several genetic conditions, in which olfactory system malformation are common and require extensive workup. The current data about the olfactory nerve, bulb and tract are important in neurosurgery and neurology since these structures are often implicated in the diseases of the nervous system. Nevertheless, taking into account the advances in the understanding of the olfactory system and its relationship with other systems, the list of implicated specialists includes otolaryngologists, craniofacial surgeons, respiratory medicine physicians, endocrinologists, geneticists and psychiatrists. Therefore, the purpose of this chapter is to sum up the current data on the clinical anatomy and developmental anomalies of olfactory nerve, bulb, tract as well as their link to other conditions.

Keywords: olfactory nerve, olfactory bulb, olfactory tract

INTRODUCTION

Olfactory sense is one of the oldest in terms of evolution senses that plays a crucial role as a way to interact with the environment. During the history of mankind, it served as an essential mechanism for identifying food, potential mating partners, dangers and enemies, therefore acting both in surveillance and pleasure mechanisms (Sarafoleanu et al. 2009, Pinto 2011). With the development of human beings, we became to some degree less dependent of olfaction, due to evolutionary, social and environmental changes (Pinto 2011). Nevertheless, it is hard to imagine even the simplest daily routine without the sense of smell.

The olfactory nerve (typically referred to as cranial nerve one) along with other neuroanatomical structures of the nasal passages is responsible for the transmission of the sense of smell. Examination of the cranial nerves is one of the fundamental parts of a complete neurological assessment of the patient. Anosmia can be the first and only sign of a lesions in basi-frontal areas (Damodaran et al. 2014). It also is a frequently reported condition that affects 3,2–22% of the population (Boesveldt et al. 2017). With the development of imaging techniques, there is an increased awareness of the possible anatomical variations and anomalies of the nervous system, which are important to consider in clinical practice. Although the information about some of the cranial nerves is abundant, cranial nerve I is relatively understudied (Covantev 2018).

From the neurosurgical point of view, the specific location of the olfactory bulbs and tracts in the anterior cranial fossa in a narrow and complicated area makes them vulnerable to injury from traumatic or pathological processes (López-Elizalde et al. 2018). Moreover, damage to the olfactory bulb and tract is a frequently described neurosurgical complication of surgery in the frontal region. Neurosurgeons personally see a large array of anatomical variations of the olfactory neve, bulb and tract. These particularities besides being an intraoperative surprise often impact the surgical procedure.

At first glance, it seems that variations, anomalies and particularities of the olfactory nerve, bulb and tract are important primarily for neurologists and neurosurgeons. Nevertheless, taking into account the advances in the understanding of the olfactory system and its relationship with other systems, the list of implicated specialists extends greatly and includes otolaryngologists, craniofacial surgeons, respiratory medicine physicians, endocrinologists, geneticists and psychiatrists.

THE NASAL CAVITY

The anatomy of the olfactory system would be senseless without a brief overview of the nasal cavity. The nose is designed to facilitate the movement of inspired air toward the olfactory epithelium.

The nasal cavities are paired and separated by the nasal septum. Each cavity is limited anteriorly by the external nares, laterally by the conchae, medially by the nasal septum, posteriorly by posterior nasal aperture, superiorly by the cribriform plate of the ethmoid bone and inferiorly by the hard palate. The lateral wall of the nasal cavity is formed by the maxilla, the perpendicular plate of the palatine bone, the medial pterygoid plate of the sphenoid bone, the labyrinth of the ethmoid and the inferior concha.

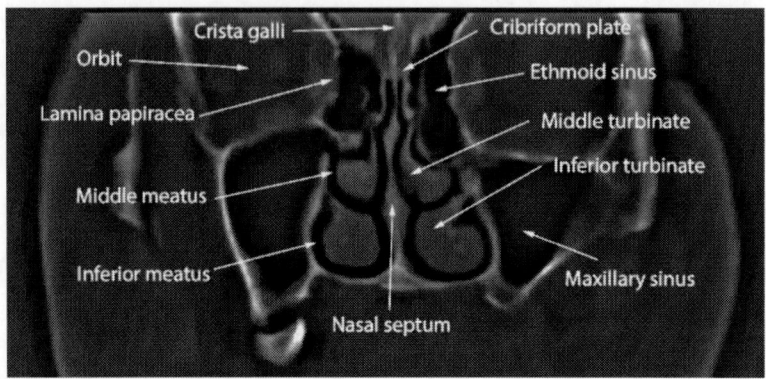

Figure 1. CT of the nasal cavity (coronal section).

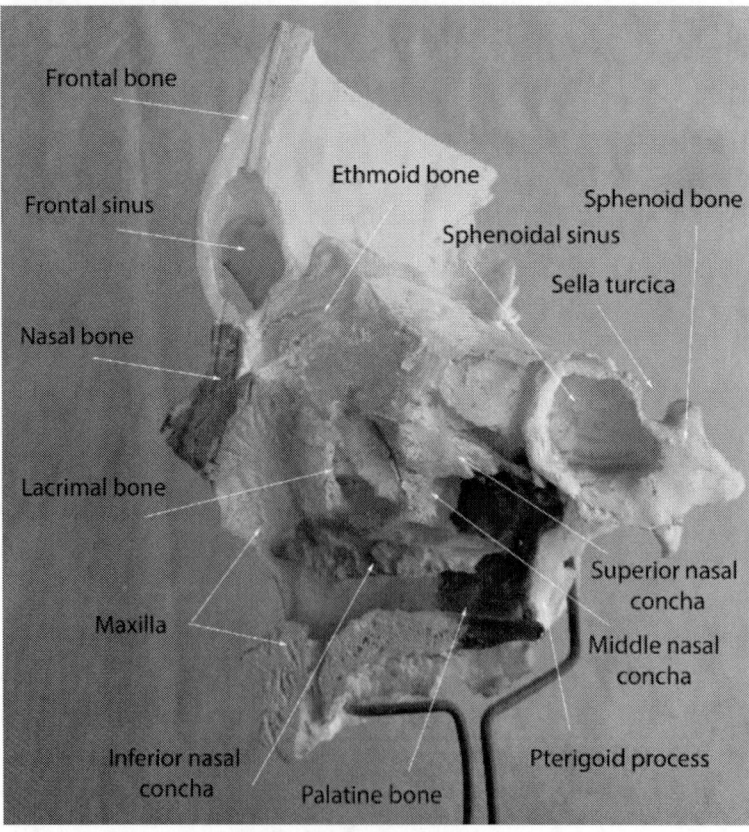

Figure 2. Bones of the nasal cavity (sagittal section).

The roof of the nasal cavity is formed by the nasal bones, frontal bones, the cribriform plate of the ethmoid and the body of the sphenoid. The floor of the nasal cavity is formed by the palatine process of the maxilla anteriorly and the horizontal plate of the palatine bone posteriorly. Finally, the medial wall is formed by the nasal septum. It is formed by the perpendicular plate of the ethmoid bone anteriorly and the vomer posteriorly (Figures 1 and 2).The nasal cavity is divided into two main segments: the respiratory and olfactory (Figure 3). The most of the nasal cavity is lined by the respiratory epithelium (ciliated pseudostratified columnar epithelium).

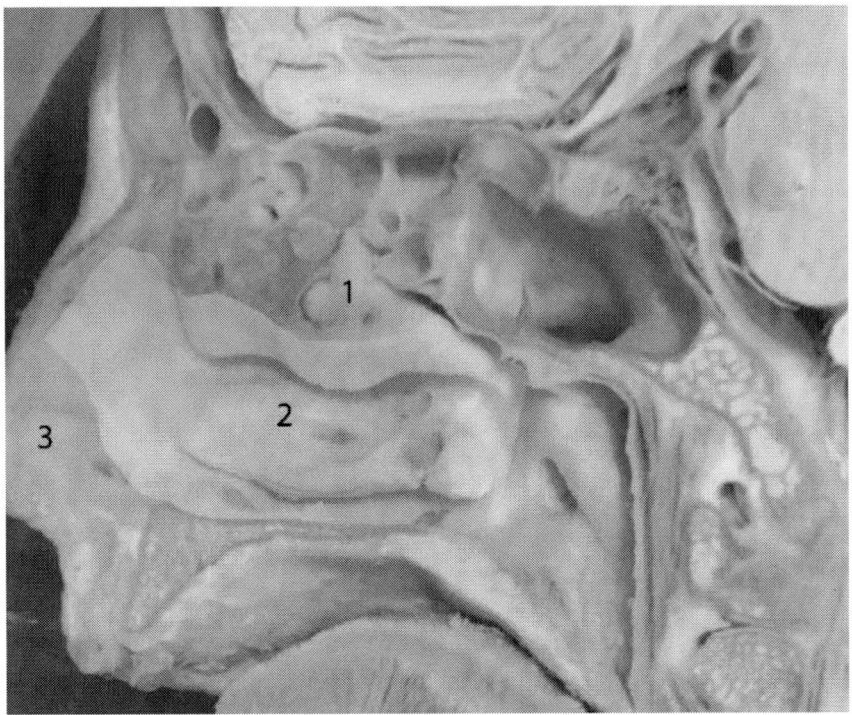

Figure 3. Main regions of the nasal cavity according to their functions (sagittal section). 1 – olfactory epithelium, 2 – respiratory epithelium, 3 – scuamous epithelium.

The remaining part of the nasal cavity is lined with olfactory epithelium (pseudostratified columnar epithelium) that has receptors for smell. The main cell types of olfactory mucosa are basal stem cells, mesenchymal stem cells, olfactory ensheathing cells and support cells (Figure 4) (Alvites et al. 2017). The olfactory neurons are sensory cells specialized for detecting odorants. Sustentacular cells maintain homeostasis and proliferation of olfactory neurons. Bowman glands are tubuloalveolar structures that allow odorant diffusion to sensory receptors. Basal cells provide replacement of olfactory neurons and sustentacular cells. Olfactory ensheathing cells also known as olfactory Schwann cells as they provide myelination for axons of olfactory neurons and assist axonal regeneration (Alvites et al. 2017). The most frequently used classification of the congenital nasal deformities was proposed by Losee and coworkers and divides the anomalies into four categories (Losee et al. 2004):

- Type I - Hypoplasia and atrophy (paucity, atrophy, or underdevelopments of skin, subcutaneous tissue, muscle, cartilage, bone)
- Type II - Hyperplasia and duplications (anomalies of excess tissue, ranging from duplications of parts to complete multiples)
- Type III - Clefts (based on Tessier classification of craniofacial clefts)
- Type IV - Neoplasms and vascular anomalies (benign and malignant neoplasms)

Type I and type III are more important for the current topic as they can sometimes be associated with olfactory system malformation. An example of a type I anomaly is congenital absence of the nose (arhinia), which can be total (total absence of the nose and rhinencephalon) and partial (partial absence of the nose). Both are associated with somatic anomalies in 50% of cases. The most common associated anomalies are absence of olfactory bulbs and nerves, absent paranasal sinuses, high arched or cleft palate, eye anomalies, low set ears and central nervous system abnormalities (Cohen and Goitein 1987). Arhinia appears due to maldevelopment of nasal placode on one or both sides leading to growth failure of medial and lateral nasal prominences. This also leads to developmental failure of olfactory bulb and tract (Olsen Ø et al. 2001). Arhinia is an extremely rare anomaly and there have been only incidental case reports (less than 60 cases). This anomaly seems to be sporadic, although some of the cases are due to genetic mutation (Mondal and Prasad 2016).

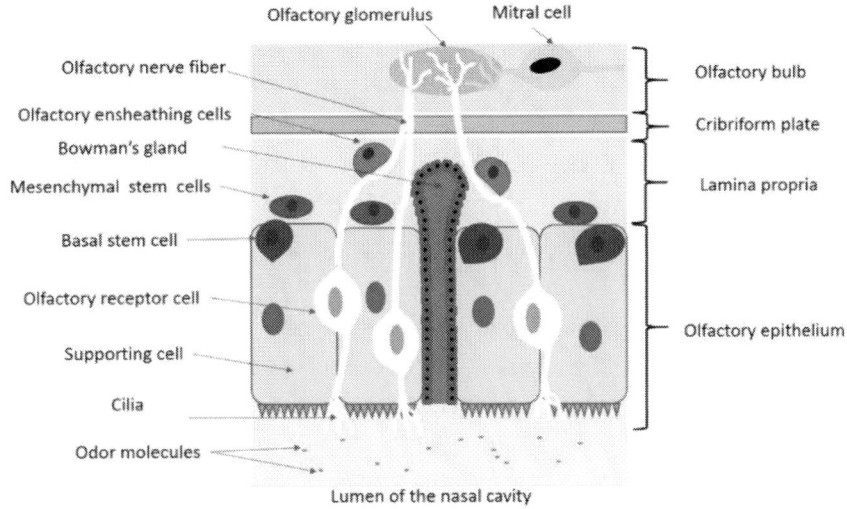

Figure 4. Main cell types of the olfactory mucosa.

Another example of a type I anomaly is congenital nasal pyriform aperture stenosis, which results from bony overgrowth in the nasal process of the maxilla, with a normal-shaped palate or alternatively deficiency of the primary palate, associated with a triangular hard palate (Ey et al. 1988). Congenital nasal pyriform aperture stenosis can be an isolated anomaly or associated with other defects (as a part of holoprosencephaly) (Tavin, Stecker, and Marion 1994). The list of associated anomalies includes microcephaly, hypoplasia of the corpus callosum or olfactory bulbs, facial anomalies and pituitary dysfunction (Captier et al. 2004). Several genetic conditions are also sometimes associated with congenital nasal pyriform aperture stenosis such as RHYNS syndrome (retinitis pigmentosa, hypopituitarism, nephronophthisis, and skeletal dysplasia), VACTERL syndrome (vertebral anomalies, anorectal malformations, cardiovascular anomalies, tracheoesophageal fistula, esophageal atresia, renal and/or radial anomalies, limb defects) and CHARGE syndrome (coloboma of the eye, heart defects, atresia of the nasal choanae, retardation of growth and/or development, genital and/or urinary abnormalities, and ear abnormalities/deafness) (Guilmin-Crépon et al. 2006). CHARGE syndrome is frequently associated with absence or hypoplasia of the olfactory bulbs and olfactory sulci (Blustajn et al. 2008).

Another type I anomaly is congenital choanal atresia which represents complete obstruction of the posterior nasal apertures (choanae) solely by osseous tissue or in combination with non-osseous tissue. The two main skeletal deformities in choanal atresia are medialization of the medial pterygoid plates and thickening of the posterior vomer. The incidence is 1:5.000-8.000 births and females are more frequently affected (female/male ratio 2:1) and unilateral defects are more common than bilateral. There are several explanation of the embryological basis of this anomaly (Schwartz and Savetsky 1986, Keller and Kacker 2000). It may be the result of persistent buccopharyngeal membrane from the foregut, persistent or abnormally located mesoderm that forms adhesions in the nasochoanal region, persistent nasobuccal membrane of Hochstetter or abnormal neural crest cell migration and mesoderm flow (Lesciotto et al. 2018, Kwong 2015, Corrales and Koltai 2009). Choanal atresia may be ab isolated anomaly or part of CHARGE syndrome (in 20%-50% of cases) (Tellier et al. 1998).

Orofacial clefts represent a type III anomaly and although the olfactory system is morphologically intact its function is often altered. Their incidence is approximately 1 in 700 live births, making them one of the most common birth defects (Dixon et al. 2011). Several studies demonstrated that orofacial clefts can lead to some degree of olfactory deficit (Richman et al. 1988, Grossmann et al. 2005). Interestingly, this relationship is also true for parents of patients with orofacial clefts. Olfactory deficit is more common in unaffected parents compared with baseline controls (41.7% versus 12.6%). This relationship is valid both for fathers (41.7% vs 12.6%) and mothers (41.7% vs 10.4%). Therefore, children with orofacial cleft who have olfactory deficit after surgery may reflect a phenotype of the disease instead of secondary consequences of the surgery (May et al. 2015). Moreover, type III anomalies can be part of other syndromes where there is olfactory deficit due to olfactory system malformations such as DiGeorge and Kallmann syndromes (Sobin et al. 2006, Hardelin and Dodé 2008).

THE OLFACTORY NERVE

What is usually referred to as the olfactory nerve is actually a group of nerves which along with the olfactory tract and bulb are an outgrowth of the forebrain (Monkhouse 2005). The upper portion of the superior turbinate, lateral surface of the posterosuperior portion of the nasal cavity and the septum is covered by the olfactory epithelium. This epithelium contains cell bodies of bipolar olfactory neurons. The dendrites of these cells are directed towards the lumen and interact with odor particles. The olfactory neurons group and extend apically thus forming olfactory nerves. The olfactory nerves are predominantly located on the roof of the nasal cavity and are represented by two main groups. The first group is the lateral olfactory nerves (12–20) which can be found at the level of the superior nasal concha. The second group is the medial olfactory nerves (12–16) which descend along the nasal septum (Barral and Croibier 2009). Therefore, each nostril has 10-20 million olfactory nerve fibers at a narrow surface area of 2-3 cm^2 (Pinto 2011). They unite and form approximately 20 filaments, which pass through the olfactory foramena of the cribriform plate, penetrate the dura mater and end up in the olfactory bulb (Figure 5). The cribridorm plate has 40-50 holes with a diameter less than 1 mm, through which arteries, veins, connecting tissues, and nerve fibers traverse (Figures 6, 7) (Habu et al. 2009).

Isolated olfactory nerve agenesis is an extremely rare anomaly which can cause anosmia (Carswell et al. 2008). It is more often associated with agenesis or hypoplasia of olfactory bulb and tract as well as other malformations. Therefore, isolated absence of one nerve should be further evaluated to exclude other possible anomalies. Damage to the olfactory nerves leads to unilateral anosmia or in less severe cases hyposmia (miscrosmia).

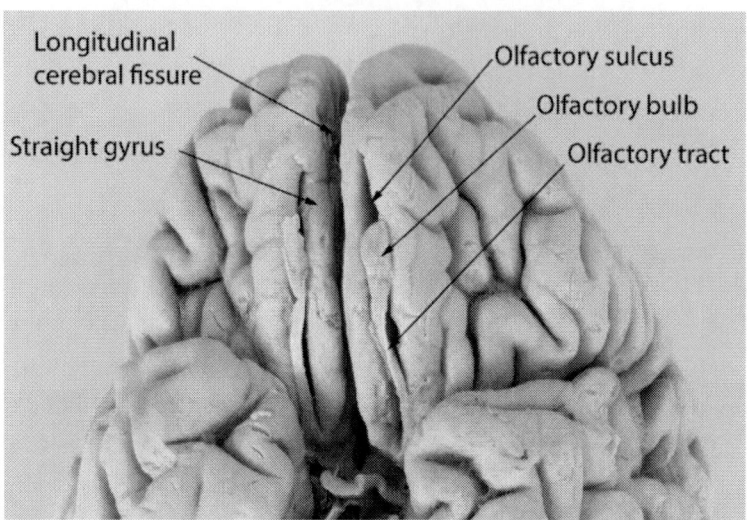

Figure 5. Olfactory bulb and tract.

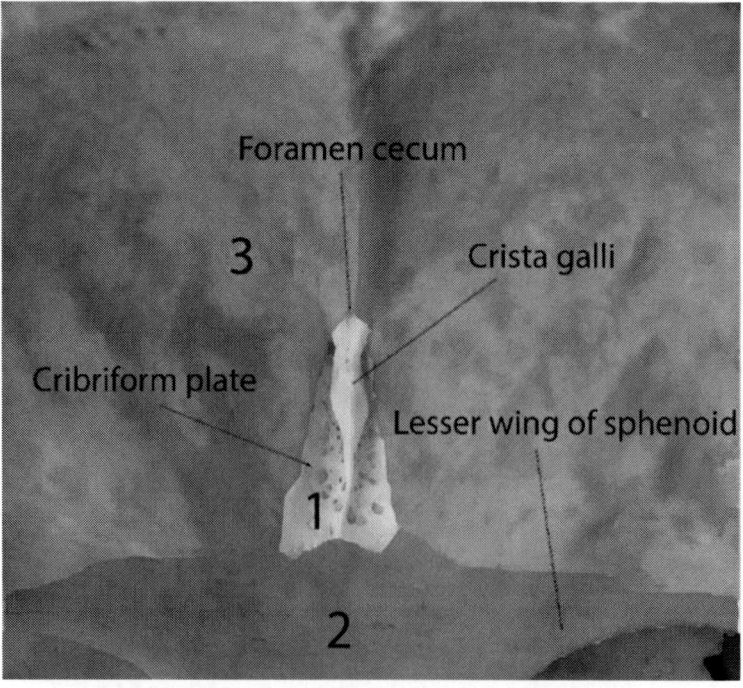

Figure 6. Anterior cranial fossa. 1 – ethmoid bone, 2 – sphenoid bone, 3 – frontal bone.

Clinical Anatomy of the Olfactory Nerve, Bulb and Tract

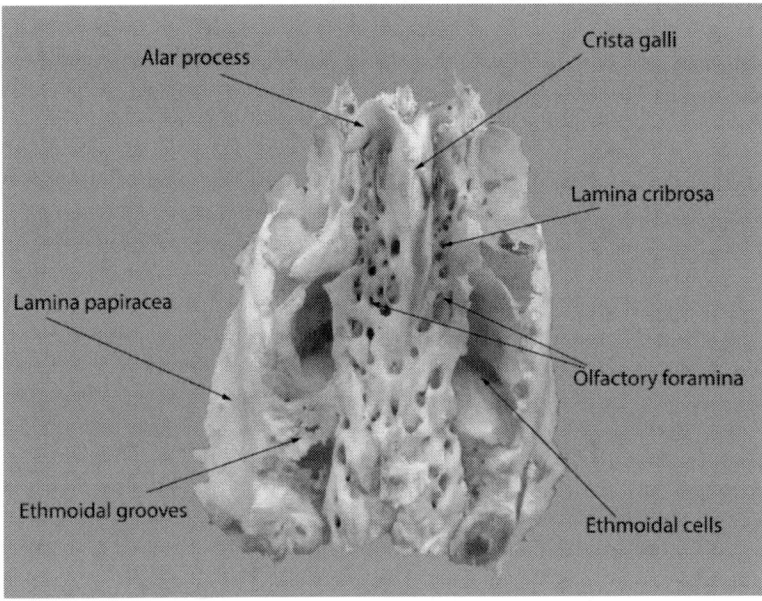

Figure 7. The ethmoid bone.

THE OLFACTORY BULB

The olfactory bulb is an ovoid structure located on the inferior surface of the frontal lobe between the straight gyrus and the medial orbital gyrus and has average length 12 mm and width 5 mm (Figure 5). It lies in the olfactory fossa within the anterior cranial fossa above the orbital plate of the frontal bone (Figure 6). The olfactory fossa is a small depression in the anterior cranial fossa and is formed by the cribriform plate of the ethmoid bone (Vaid and Vaid 2015). This thin and perforated bone is a boundary that isolates the nasal cavity from the anterior cranial fossa (Figure 7). Since the lateral laminae of the cribriform plate are the thinnest portions of the ethmoid roof, it is a particularly vulnerable area. In some cases, increased intracranial pressure can result in damage and cerebrospinal fluid leak or meningitis (Rai et al. 2018).

The medial boundary of the olfactory fossa is formed by the crista galli and the lateral limit by the lateral lamella of the cribriform plate (Jacob and

Kaul 2014). This zone is vulnerable in case of congenital defects of bone development.

In some cases, defects or zones of small resistances of the ethmoid bone are accompanied with herniation of the structures located in the cranial cavity. The contents may include the meninges (meningocele), meninges and brain (meningoencephalocele), or a part of ventricle (hydroencephalomeningocele). Herniation at the level of the cribriform plate is known as a nasal encephalocele, which can be divided into basal (40% of cases) and sincipital (frontoethmoidal) (60% of cases). Basal encephaloceles are classified into (Velho et al. 2019, Rai et al. 2018):

1. transethmoidal - a sac protruding through a defect in the cribriform plate into the superior meatus and appearing as a nasal polyp,
2. sphenoethmoidal - a sac herniating through the cribriform plate between the posterior ethmoidal cells and sphenoidal sinus into the nasopharynx,
3. spheno-orbital - a sac protruding through the superior orbital fissure into the orbit, and lastly,
4. transsphenoidal - a defect in the posterior cribriform plate leading to a sac herniating into the nasopharynx

Frontoethmoidal encephaloceles present as a soft, compressible, external mass over the glabella and are subtyped into:

1. nasoethmoidal (most common) - site of herniation at the level of foramen cecum, with the mass located at the nasal bridge,
2. nasofrontal – site of herniation at the level of fonticulus nasofrontalis, with the mass located at forehead or nasal bridge,
3. naso-orbital (least common) - site of herniation at the level of median orbital wall with the mass located in the orbit (Velho et al. 2019).

In rare cases encephalocele can be associated with the olfactory bulb or nerves abnormalities, particularly in cases of bone defects or tumors (Habu et al. 2009).

There a several types and subtypes of neurons in the olfactory bulb. They have been categorized based on the layers in which their cell bodies are found (Figure 8). The axons of olfactory sensory neurons make synapses in the glomerular layer, each for a single odorant. Neurons surrounding glomeruli in the glomerular layer are periglomerular cells (send inhibitory synapses to mitral and tufted cells), external tufted cells and superficial short-axon cells. All of the periglomerular cells are interneurons that help coordinate glomerular output. The two main types of projection neurons are the mitral cells and the tufted cells, which send their axons to the olfactory cortex.

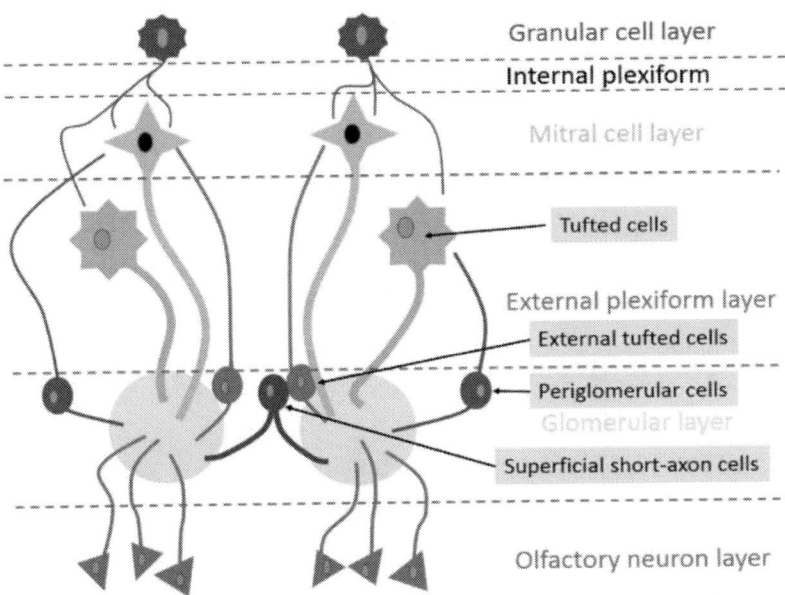

Figure 8. The olfactory bulb network.

The cell body of mitral cells are located in the mitral cell layer. The tufted cells are scattered throughout the external plexiform layer. The internal plexiform layer consists of axons of the mitral cells, external tufted cells. The granule cell layer consists of granule cells, which are axon-less interneurons extending dendrites apically into the external plexiform layer (Monkhouse 2005, Leboucq et al. 2013, Cardali et al. 2005, Nagayama, Homma, and Imamura 2014). Therefore, all impulses from the olfactory epithelium are transmitted to the olfactory bulbs and from there to the primary olfactory cortex. Severe damage to the olfactory bulbs interrupts neuronal transmission leading to anosmia and gustatory changes.

Levy and coworkers demonstrated in their study of 220 patients who undergone MRI that the olfactory bulbs are bifurcated 5% of cases and trifurcated in 0,45%. Olfactory sulcal depth was similar in all patients but olfactory bulbs were smaller in patients with supernumeral olfactory bulbs compared to subjects with a single bulb. In all patients with supernumeral olfactory bulbs the olfactory grooves were abnormal: widened, flat or absent. The bifurcation of the olfactory bulbs doesn't seem to relate to olfactory function although one of the patients had congenital loss of smell (Levy et al. 2012). Interestingly, some animals have vomeronasal (Jacobson's) organ and receive sensory input with the help of the accessory olfactory (Scalia and Winans 1975). Therefore, the presence of an accessory olfactory bulb in some patients may be a reminder of an evolutionary remnant.

Aplasia or hypoplasia of the olfactory bulbs can be isolated or associated with other malformations (Rienzo, Artuso, and Colosimo 2002). Isolated anomaly is rare and can affect only one olfactory bulb. There are three morphological types of the anomaly (Yousem et al. 1993, Truwit et al. 1993, Knorr et al. 1993):

- Type I - hypoplasia of the olfactory bulbs with olfactory tracts present;
- Type II - aplasia of the olfactory bulbs with olfactory tracts present;
- Type III - aplasia of both olfactory bulbs and olfactory tracts.

The extent of associated anomalies varies. In some cases, the anomalies are of little clinical significance such as bilobed pineal body

(Petty 1965). In other cases, the malformations are extensive and affect the nervous, cardiovascular, gastrointestinal and musculoskeletal systems (Stewart 1939, Morton 1947).

Anomalies of the olfactory system are probably underestimated. Patients with abnormal ciliary motion and congenital heart disease have right olfactory bulb dysmorphometry (hypoplasia or aplasia) in 21.4% of cases, left olfactory bulb dysmorphometry in 14.3% and frequent olfactory sulci abnormalities (Panigrahy et al. 2016).

Interestingly, for a long period of time the presence of olfactory bulbs was considered essential for the sense of smell. Nevertheless, Weiss and coworkers demonstrated that in 0.6% of patients there is no anatomically defined olfactory bulbs on MRI without anosmia. Therefore, in some cases extreme plasticity allows preservation of olfactory function even in the presence of anatomical abnormalities (Weiss et al. 2020).

Olfactory bulb cysts or ventricles are incidental findings (Roy, Frost, and Schochet 1987). In some species there is an embryologic cavity inside the olfactory bulbs which is termed olfactory bulb ventricles. Therefore, the same embryological remnant can be seen in humans. Their incidence is debatable and ranges from 5.5% to 59% depending on the method (Smitka et al. 2009, Burmeister et al. 2011).

Although olfactory bulb malformations can be an isolated anomaly or an anomaly of little clinical relevance the patient should be evaluated for other malformations and consulted by a geneticist. Aplasia or hypoplasia of the olfactory bulbs is one of the main findings is Kallmann's syndrome. Kallmann's syndrome is a genetic condition, that appears due to a mutation in the KAL1 (ANOS1) gene. There are also other genes, the mutation in which causes this condition like FGFR1, PROKR2, PROK2, CHD7 and other (Dodé and Hardelin 2009). The other particularly important feature is hypogonadotropic hypogonadism. This is due to the failure of cells that normally express luteinizing hormone-releasing hormone to migrate from the medial olfactory placode into the forebrain (Truwit et al. 1993). There is also failed neuronal migration from the lateral olfactory placode along the olfactory fila to the forebrain which leads to aplasia or hypoplasia of the olfactory bulbs and tracts (Truwit et al. 1993). Loss of smell in

Kallmann's syndrome us due to aplasia of the olfactory bulb and tract in 68-84% and hypoplasia in 16-32% (Yousem et al. 1996). This is not the only central nervous anomaly and patients frequently have other structural defects like volume loss of the frontal and temporal lobes, corpus callosum defects, arachnoid cyst, empty sella turcica, pituitary hypoplasia etc (Yousem et al. 1996, McCabe et al. 2011, Massimi et al. 2016, Takahashi et al. 1997, Madan et al. 2004). There can also be craniofacial malformations, altered bone morphology and other system anomalies. There a significant phenotypical overlaps with other conditions like CHARGE-syndrome, septo-optic dysplasia and combined pituitary hormone deficiency.

Malformations of the olfactory bulb and/or tract are also reported in other syndromes although they do not represent the main feature of these genetic conditions. These conditions include Waardenburg syndrome, Meckel-Gruber syndrome, Fryns syndrome, short rib-polydactyly syndrome, Jacobsen syndrome, Johanson-Blizzard syndrome, DiGeorge syndrome, trisomy 13, trisomy 18 and other (Pingault et al. 2013, Paetau, Salonen, and Haltia 1985, Van Hove et al. 1995, al-Gazali et al. 1999, Booth and Rollins 2016, Aziz 1980).

THE OLFACTORY TRACT

The axonal projections of mitral and tufted cells form bundles that merge and run posteriorly along the olfactory sulcus forming the olfactory tract. The tract ends up with a triangular widening called the olfactory trigone, which is located superior to the anterior clinoid process and rostral to the anterior perforated substance. At this point the tract divides into the lateral, medial and intermediate olfactory stria (Figure 9) (Binder, Sonne, and Fischbein 2010).

The lateral stria consists of axons that run to the primary olfactory cortex, located within the uncus of temporal lobe. It connects with the anterior olfactory nucleus which is located between the olfactory bulb and tract. The anterior olfactory nucleus receives fibers from the tufted cells and sends axons to the contralateral anterior olfactory nucleus and olfactory bulb (via anterior commissure) or the ipsilateral olfactory cortical areas. The primary olfactory cortex includes the piriform cortex and periamygdaloid complex. Fibers from the primary olfactory cortex run to the entorhinal cortex (secondary olfactory cortex) to the hippocampus. Other fibers end up in the insula, frontal lobe by way of the uncinate fasciculus; amygdala, lateral preoptic hypothalamus, and the nucleus of the diagonal band; mediodorsal thalamic nucleus to the orbitofrontal cortex (for conscious analysis of odor, a phylogenetically newer pathway) (Binder, Sonne, and Fischbein 2010, Doty and Bromley 2007).

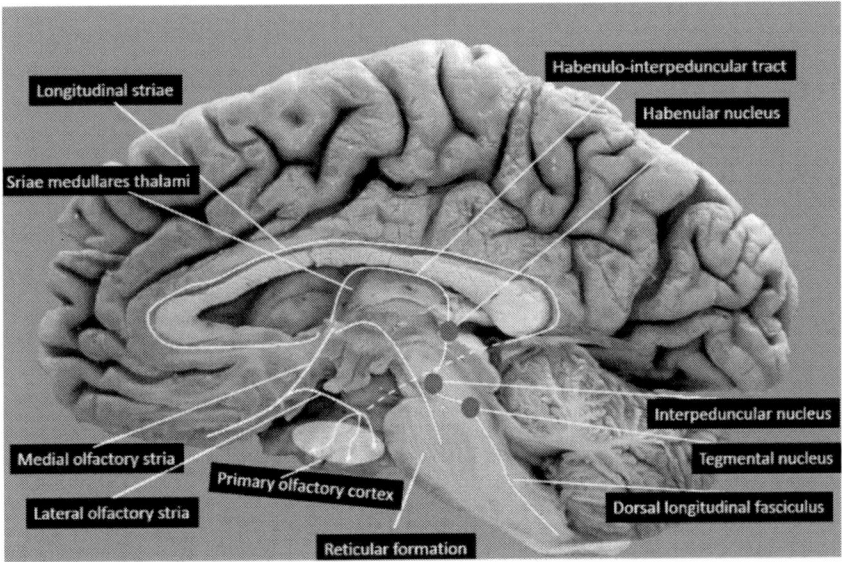

Figure 9. Olfactory tract.

The medial stria consists of axons that across the medial plane of the anterior commissure, where they meet the olfactory bulb of the opposite side. It also projects to the septal area (subcallosal area and the paraterminal gyrus) known as the medial olfactory area. This phylogenetically older pathway mediates the emotional/autonomic response to odors with its limbic connections (Binder, Sonne, and Fischbein 2010).

In cases when there is the intermediate (anterior) olfactory stria it runs to the anterior perforated substance to an area of gray matter, which is called the olfactory tubercle. The fibers are directed to the anterior commissure and travel to the contralateral olfactory cortex (Binder, Hirokawa, and Windhorst 2009).

ARTERIAL SUPPLY AND VENOUS DRAINAGE

The olfactory nerve has two main sources of arterial supply: intracranial and extracranial. The intracranial sources are olfactory artery, accessory olfactory artery, anterior and posterior ethmoidal arteries. The olfactory artery and accessory olfactory artery arise from anterior cerebral artery or anterior communicating artery. The anterior and posterior ethmoidal arteries arise from ophthalmic artery. The extracranial vascular supply is by sphenopalatine artery, anterior and posterior ethmoidal arteries (Figure 10). The sphenopalatine artery arises from the external carotid artery (Hendrix et al. 2014).

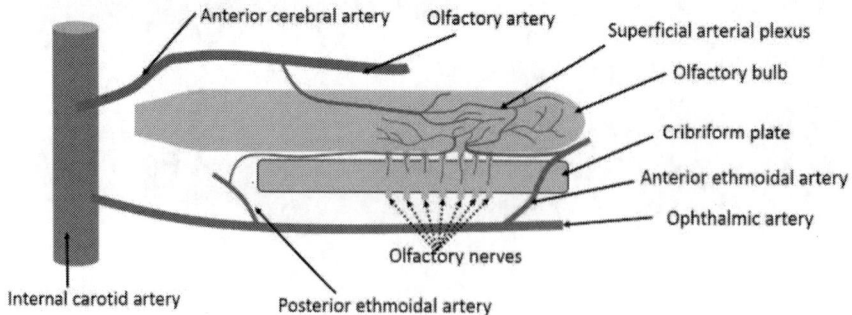

Figure 10. Arterial supply of the olfactory bulb and nerves.

The major source of the arterial supply of the olfactory bulb and tract is the olfactory artery. In over half of the cases it originates from the lateral surface of the anterior cerebral artery (A2 segment). In the rest of the cases it originates from the medial frontobasal artery, which in turn is a branch of the anterior cerebral artery. It provides supply in front of the olfactory trigone in the medial olfactory sulcus, the olfactory tract and olfactory bulb with a maximum of three terminal branches (Favre et al. 1995, Tsutsumi, Ono, and Yasumoto 2017).

The olfactory tract receives arterial blood supply from various sources, which include posterior orbitofrontal artery, medial orbitofrontal artery, posterior orbitofrontal artery and medial striate artery (Ciołkowski, Michalik, and Ciszek 2004).

In rare cases an artery originates from the terminal portion of internal carotid artery which is called persistent primitive olfactory artery. It is encountered in 0.14-0.64% making it a rare developmental variation (Uchino et al. 2011, Vasović et al. 2013). There are three types of this variation (Sato et al. 2015, Horie et al. 2012):

- Type 1 – runs along the olfactory tract and makes a hairpin turn to supply the territories of the distal anterior cerebral artery;
- Type 2 - extending to the cribriform plate to supply the nasal cavity;
- Type 3 (transitory type, two branches) - a branch making a hairpin turn and supplying the distal part of the anterior carotid artery territory and a branch extending to the cribriform plate to supply the nasal cavity.

There are several reports of aneurysms associated with persistent primitive olfactory artery particularly located at the characteristic hairpin bend (Sato et al. 2015, Horie et al. 2012, Yamamoto et al. 2009, Tsuji, Abe, and Tabuchi 1995).

There are several vascular malformations that are important to consider in the differential diagnosis of olfactory dysfunction. Aneurysms of the primitive olfactory artery, arteriovenous fistulae involving the anterior ethmoidal arteries and pial-dural arteriovenous malformations that involve the primitive olfactory artery can cause changes in olfaction due to compression of the olfactory bulb or tract (Jimbo et al. 2010, Inoue et al. 2014, Yamamoto et al. 2009).

Olfactory bulb or tract ischemia which results in anosmia or hyposmia are generally rare. This is due to a vast network of arteria supply with multiple anastomosis. Therefore, damage to the olfactory system usually requires occlusion of multiple branches or of the main supplying artery (Zhan et al. 2007). Alternatively, olfactory bulbs themselves have a complex network that ensures the correct processing of the olfactory inputs. They also have high degree of plasticity and neurogenesis that allows to withstand such damage (Díaz et al. 2017).

Olfactory areas of the brain are widespread, difficult to localize and also vary from person to person. The primary olfactory cortices are mostly supplied by the lenticulostriate arteries arising from the first segment of the middle cerebral arteries and by the posterior temporal branches of the middle cerebral artery (Duvernoy et al. 1991). Therefore, an isolated unilateral stroke in the primary olfactory area usually will not lead to olfactory symptoms. The reason for this is the contralateral area that continues to receive bilateral olfactory information. Nevertheless, there are rare cases that demonstrate that the olfactory cortex shares some common vascular supply with the gustatory cortex, and abnormalities of taste perception may develop with isolated strokes of the insula (Cereda et al. 2002). For example, left posterior insula lesion may affect taste and olfactory perception similarly by increasing sensitivity contralateral to the lesion (Mak et al. 2005).

The venous system of the olfactory system is relatively understudies compared with the arterial system. The main sources of venous drainage are branches of the olfactory vein, the frontopolar vein, the orbital vein, and small affluent branches of the sagittal sinus (Wang et al. 2008).

SURGICAL CONSIDERATIONS

Anatomical dissection is an essential method to study anatomy of a particular region, especially neuroanatomy. Nevertheless, this knowledge, although is important but differs from the anatomy seen by the eyes of a surgeon. During surgery it is often hard or unnecessary to perform rigorous dissection of the anatomical structures. This is also true for the olfactory system. Damage to the olfactory bulb and tract is a frequently described complication of brain surgery in the frontal region (Cömert et al. 2011). Surgeons mention that the circumstances in which they find the olfactory neve, bulb and tract differ. The difference is anatomical (length and thickness of the olfactory nerve) as well as biomechanical (resistance against compression and tension). This is also true for the olfactory bulbs which in some cases are adherent to the olfactory fossa or on the opposite easily detachable and mobile (Dietz 1981). Dissecting the olfactory nerve from the orbitofrontal cortex can be performed during operations without its damage but bilateral protection is not always possible (Suzuki, Mizoi, and Yoshimoto 1986). Mobilization of the olfactory bulb and tract is limited to approximately 29.3±6.4 mm in length, whereas frontal lobe retraction is limited to 10-15 mm otherwise there is also risk for compromising the olfactory function (Cardali et al. 2005, Cömert et al. 2011). Dissection near the olfactory artery and medial orbitofrontal arteries, which are located underneath the olfactory tract in the olfactory sulcus should be done carefully due to possible damage and subsequent bleeding during surgery (Wang et al. 2008, Cömert et al. 2011).

Nowadays, functional endoscopic sinus surgery is a common modality of treatment for diseases of nose and paranasal sinuses. The most known complications encountered during sinus surgeries are skull base entry, orbital and ocular injuries (McMains 2008). The olfactory fossa where the olfactory bulb and tract are located is protected by a thin and perforated ethmoid bone. Therefore, there is risk of iatrogenic trauma to the structures of the anterior cranial fossa particularly at the level of the olfactory fossa. In 1962 Keros proposed a classification based on examination of 450 skulls and analysis of the relation between the olfactory fossa and the ethmoid roof. The depth of was measured by the vertical height of the lateral lamella of the cribriform plate, the difference between the height of the cribriform plate and ethmoid roof. As a result, Keros proposed three categories (Keros 1962):

- Type I - the depth is 1 to 3 mm, with short lateral lamella, and the ethmoid roof is almost in the same horizontal plane as the cribriform plate of ethmoid
- Type II - the depth is from 4 to 7, with longer lateral lamella
- Type III - the depth is 8 to 16 mm, and the ethmoid roof lies essentially higher than the cribriform plate (Figure 11).

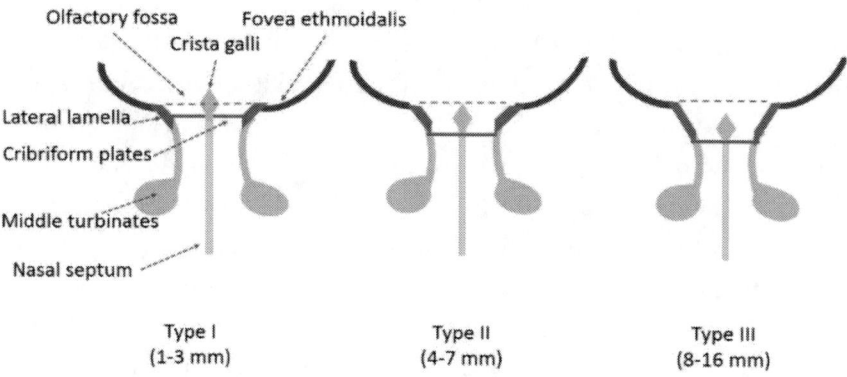

Figure 11. Schematic representation of the Keros classification.

Based on this classification, the height of the lateral lamella of the cribriform plate is proportional to the risk of its iatrogenic damage. Keros type III is therefore particularly vulnerable to possible injury during surgery, trauma or tumor erosion. Keros type I is seen in 12.5-26.3% of population, type II in 53.2-73.3% and type III in 0.5-11.7% (Madani, El-mardi, and El-Din 2020, Keros 1962, Ulualp 2008).

Nowadays there are multiple interventions that directly or indirectly can compromise olfaction. Olfactory function is often affected in case of nasal surgery and endonasal procedures (Shemshadi et al. 2008). Unfortunately, it is often considered as secondary among the sensory functions, which reflects a lack of interest in sparing olfaction after surgery (Jang et al. 2013). Surgical excision of olfactory epithelium is sometimes performed to relieve phantosmia while preserving olfaction (Leopold, Loehrl, and Schwob 2002). *Tumors* arising from the olfactory mucosa, olfactory fibers and anterior cranial fossa can also cause loss of *smell*.

Olfactory neuroblastoma (also known as esthesioneuroblastoma) is a very rare tumor that develops in the upper part of the nasal cavity and arises from neural tissue associated with olfaction (Thompson 2009). Olfactory meningiomas also can cause anosmia or hyposmia and this is often the first symptom of the disease. Moreover, olfaction is often hard to preserve in olfactory groove meningiomas particularly if the tumor's size is more than 4 cm (Jang et al. 2013). The trans-sphenoidal approach for pituitary adenoma surgery in some cases can affect the neighboring structures. The olfactory bulbs can decrease one year after surgery and this correlates with loss of smell (Podlesek et al. 2019).

One of the most classical syndromes associated with olfactory function is the Foster-Kennedy syndrome, which includes changes as optic atrophy in the ipsilateral eye, disc edema in the contralateral eye, central scotoma in the ipsilateral eye, anosmia ipsilaterally, headache and nausea. The syndrome arises due to optic nerve compression, olfactory nerve compression, and increased intracranial pressure secondary to a mass (Massey and Schoenberg 1984).

CLINICAL IMPLICATIONS

Abnormalities of olfaction include quantitative and qualitative smelling disorders (table 1) (Hummel, Landis, and Hüttenbrink 2011).

The main causes of smell disorders are trauma, viral infections, nasal disorders and neurological illnesses (Hummel, Landis, and Hüttenbrink 2011). Damage to the olfactory epithelium, the olfactory nerves, the olfactory bulb or olfactory tract leads to unilateral anosmia or in less severe cases hyposmia (miscrosmia). On the other hand, destruction of olfactory cortex or olfactory pathways posterior to the olfactory trigone does not affect olfactory function. Only bilateral damage to the olfactory pathways posterior to the olfactory trigone can lead to bilateral anosmia (Figure 12).

Table 1. Olfactory abnormalities definition and classification

Type of disorder	Olfactory dysfunction	Description
Quantitative	Anosmia	The lack of ability to smell all odours. In case of specific anosmia there is inability to smell one particular odour.
	Hyposmia	Reduced ability to smell odours
	Hyperosmia	Enhanced ability to smell
Qualitative	Parosmia	Wrong perception of odours
	Phantosmia	Perception of odours in the absence of a relevant odour source

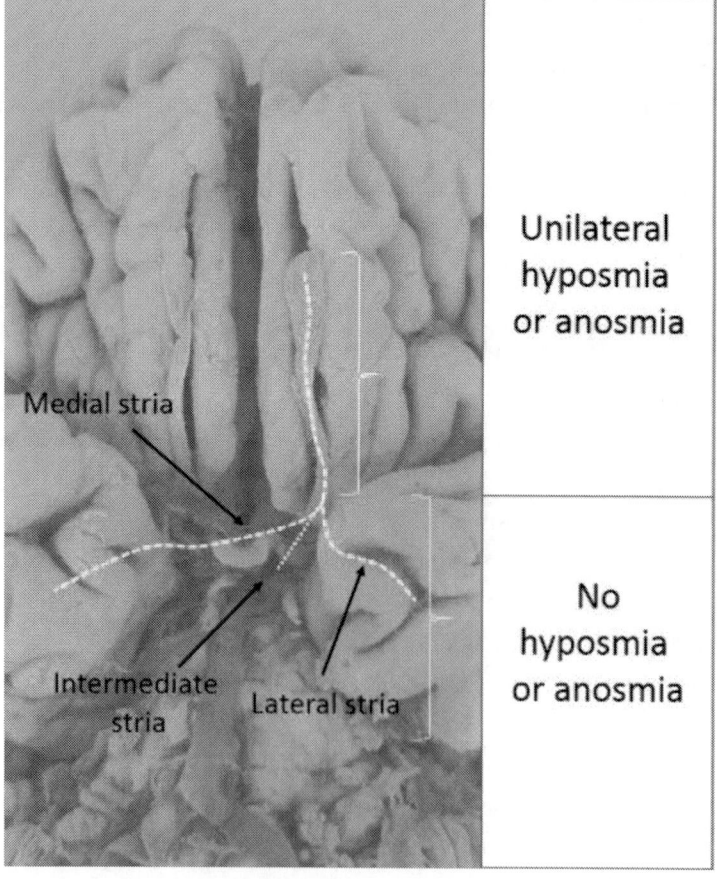

Figure 12. The level of trauma leading to anosmia.

In some cases, epileptiform activity of the olfactory cortical regions can lead to olfactory hallucinations. This can be due to a variety of causes like tumors, mesial temporal sclerosis, intracerebral hemorrhage, middle cerebral artery aneurysm, arteriovenous malformation, head injury, and encephalitis (Vaughan and Jackson 2014). Moreover, patients with temporal lobe epilepsy due to hippocampal sclerosis frequently have difficulties with odor discrimination and identification (Espinosa-Jovel et al. 2019).

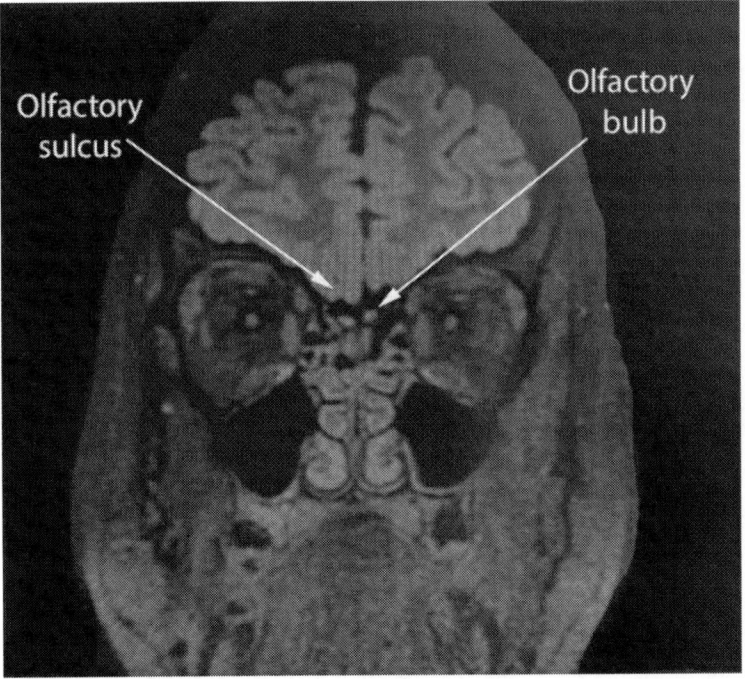

Figure 13. Assessment of the olfactory bulb and sulcus on a MRI.

The anatomy of the olfactory system cannot be studied alone since it correlates with its function. Therefore, two particular anatomical structures should be at special attention: the olfactory bulb and sulcus (Figure 13). For example, olfactory bulb volumes correlate with olfactory function (Buschhüter et al. 2008). Moreover, the difference in volume between the right and left olfactory bulbs may be partly responsible for lateralized differences in olfactory function (Hummel, Haehner, et al. 2013). The volume also tends to increase in case of early blindness and decrease in a variety of conditions, which include respiratory, neurological, psychiatric etc. (Rombaux et al. 2010, Rombaux et al. 2006). As olfactory bulbs represent a part of the central nervous system they can be a marker of several neurological or psychiatric conditions. Olfactory bulbs are smaller in patients with temporal lobe epilepsy compared to healthy controls (Hummel, Henkel, et al. 2013). The same results are observed in Parkinson's disease and schizophrenia (Li et al. 2016, Bruce I. Turetsky et al. 2000). In patients with multiple sclerosis olfactory bulb volume correlates with disease duration, attack frequency and is associated with a higher depression scores (Tanik et al. 2015, Yaldizli et al. 2016). In patients with major depression there is a significant negative correlation between olfactory bulb volume and depression scores (Negoias et al. 2010). Moreover, the volume itself can be a predictor factor of therapeutic outcome (Negoias et al. 2016).

Olfactory sulcal depth was proposed as a marker for mental illnesses (particularly schizophrenia) (Turetsky et al. 2009). Schizophrenia patients had a significantly shallower olfactory sulcus compared with the controls bilaterally (Takahashi et al. 2013). It is possible that a shallow olfactory sulcus represents abnormal forebrain development. The altered depth of the olfactory sulcus, which exists before psychosis onset, could be predictive of transition to psychosis, but also suggest ongoing changes of the sulcus morphology during the course of the illness (Takahashi et al. 2014).

Conclusion

The anatomy of the olfactory system is complicated and one of the oldest. The chapter demonstrates that the structure of olfactory nerve, bulb and tract is tightly related to its function. Variations and abnormalities of development of this extraordinary system can be encountered by any specialist and therefore require special attention during evaluation. This also underlines the importance of proper evaluation of the sense of smell.

Therefore, preserved olfactory function is an indicator not only of an anatomically intact olfactory system but can also excludes a variety of clinically important conditions.

REFERENCES

al-Gazali, L. I., L. Sztriha, A. Dawodu, E. Varady, M. Bakir, A. Khdir, and J. Johansen. 1999. "Complex consanguinity associated with short rib-polydactyly syndrome III and congenital infection-like syndrome: a diagnostic problem in dysmorphic syndromes." *J Med Genet* no. 36 (6):461-6.

Alvites, RD, ARC Santos, ASP Varejão, and de Castro Osório ACP. 2017. "Olfactory Mucosa Mesenchymal Stem Cells and Biomaterials: A New Combination to Regenerative Therapies after Peripheral Nerve Injury." In *Mesenchymal Stem Cells-Isolation, Characterization and Applications:* . Rijeka InTech.

Aziz, M. Ashraf. 1980. "Anatomical defects in a case of trisomy 13 with a D/D translocation." *Teratology* no. 22 (2):217-227.

Barral, Jean-Pierre, and Alain Croibier. 2009. "Chapter 10 - Olfactory nerve." In *Manual Therapy for the Cranial Nerves*, edited by Jean-Pierre Barral and Alain Croibier, 59-72. Edinburgh: Churchill Livingstone.

Binder, D. K., C. Sonne, and N. J. Fischbein. 2010. *Cranial Nerves: Anatomy, Pathology, Imaging*: Thieme.

Binder, Marc D., Nobutaka Hirokawa, and Uwe Windhorst. 2009. "Olfactory Tubercle." In *Encyclopedia of Neuroscience*, edited by Marc D. Binder, Nobutaka Hirokawa and Uwe Windhorst, 3025-3025. Berlin, Heidelberg: Springer Berlin Heidelberg.

Blustajn, J., C. F. E. Kirsch, A. Panigrahy, and I. Netchine. 2008. "Olfactory Anomalies in CHARGE Syndrome: Imaging Findings of a Potential Major Diagnostic Criterion." *American Journal of Neuroradiology* no. 29 (7):1266-1269.

Boesveldt, Sanne, Elbrich M. Postma, Duncan Boak, Antje Welge-Luessen, Veronika Schöpf, Joel D. Mainland, Jeffrey Martens, John Ngai, and Valerie B. Duffy. 2017. "Anosmia-A Clinical Review." *Chemical senses* no. 42 (7):513-523.

Booth, T. N., and N. K. Rollins. 2016. "Spectrum of Clinical and Associated MR Imaging Findings in Children with Olfactory Anomalies." *American Journal of Neuroradiology* no. 37 (8):1541-1548.

Bruce I. Turetsky, M. D., Paul J. Moberg, David M. Yousem, M. D., Richard L. Doty, Steven E. Arnold, M. D., and, and Raquel E. Gur, M. D., 2000. "Reduced Olfactory Bulb Volume in Patients With Schizophrenia." *American Journal of Psychiatry* no. 157 (5):828-830.

Burmeister, H. P., T. Bitter, P. A. T. Baltzer, M. Dietzel, O. Guntinas-Lichius, H. Gudziol, and W. A. Kaiser. 2011. "Olfactory bulb ventricles as a frequent finding—a myth or reality? Evaluation using high resolution 3 Tesla magnetic resonance imaging." *Neuroscience* no. 172:547-553.

Buschhüter, D., M. Smitka, S. Puschmann, J. C. Gerber, M. Witt, N. D. Abolmaali, and T. Hummel. 2008. "Correlation between olfactory bulb volume and olfactory function." *NeuroImage* no. 42 (2):498-502.

Captier, G., S. Tourbach, M. Bigorre, M. Saguintaah, J. El Ahmar, and P. Montoya. 2004. "Anatomical consideration of the congenital nasal pyriform aperture stenosis: localized dysostosis without interorbital hypoplasia." *J Craniofac Surg* no. 15 (3):490-6.

Cardali, S., A. Romano, F. F. Angileri, A. Conti, D. La Torre, O. de Divitiis, D. d'Avella, M. Tschabitscher, and F. Tomasello. 2005. "Microsurgical anatomic features of the olfactory nerve: relevance to olfaction preservation in the pterional approach." *Neurosurgery* no. 57 (1 Suppl):17-21; discussion 17-21.

Carswell, A. J., D. Whinney, N. Hollings, and P. Flanagan. 2008. "Isolated olfactory nerve agenesis." *Br J Hosp Med (Lond)* no. 69 (8):474.

Cereda, C., J. Ghika, P. Maeder, and J. Bogousslavsky. 2002. "Strokes restricted to the insular cortex." *Neurology* no. 59 (12):1950-1955.

Ciołkowski, M., R. Michalik, and B. Ciszek. 2004. "Arteries to the proximal part of the olfactory tract." *Folia Morphol (Warsz)* no. 63 (4):455-8.

Cohen, D., and K. J. Goitein. 1987. "Arhinia revisited." *Rhinology* no. 25 (4):237-44.

Cömert, Ayhan, Hasan Çaglar Ugur, Gökmen Kahilogullar, Ela Cömert, Alaittin Elhan, and Ibrahim Tekdemir. 2011. "Microsurgical Anatomy for Intraoperative Preservation of the Olfactory Bulb and Tract." *Journal of Craniofacial Surgery* no. 22 (3).

Corrales, C. E., and P. J. Koltai. 2009. "Choanal atresia: current concepts and controversies." *Curr Opin Otolaryngol Head Neck Surg* no. 17 (6):466-70.

Covantev, S. 2018. "Cranial nerve I: clinical anatomy and beyond." *Russian Open Medical Journal* no. 7:e0101.

Damodaran, Omprakash, Elias Rizk, Julian Rodriguez, and Gabriel Lee. 2014. "Cranial nerve assessment: A concise guide to clinical examination." *Clinical Anatomy* no. 27 (1):25-30.

Díaz, D., R. Muñoz-Castañeda, C. Ávila-Zarza, J. Carretero, J. R. Alonso, and E. Weruaga. 2017. "Olfactory bulb plasticity ensures proper olfaction after severe impairment in postnatal neurogenesis." *Scientific Reports* no. 7 (1):5654.

Dietz, H. 1981. Some remarks about the olfactory nerve from the surgical point of view. *In: The Cranial Nerves.* M. Samii and P.J. Jannetta, eds. Springer, Berlin, Heidelberg.

Dixon, M. J., M. L. Marazita, T. H. Beaty, and J. C. Murray. 2011. "Cleft lip and palate: understanding genetic and environmental influences." *Nat Rev Genet* no. 12 (3):167-78.

Dodé, Catherine, and Jean-Pierre Hardelin. 2009. "Kallmann syndrome." *European Journal of Human Genetics* no. 17 (2):139-146.

Doty, Richard L., and Steven M. Bromley. 2007. "Chapter 7 - Cranial Nerve I: Olfactory Nerve." In *Textbook of Clinical Neurology (Third Edition)*, edited by Christopher G. Goetz, 99-112. Philadelphia: W. B. Saunders.

Duvernoy, H. M., P. Bourgouin, E. A. Cabanis, and J. L. Vannson. 1991. *The Human Brain: Surface, Three-Dimensional Sectional Anatomy and MRI*: Springer Vienna.

Espinosa-Jovel, Camilo, Rafael Toledano, Adolfo Jiménez-Huete, Ángel Aledo-Serrano, Irene García-Morales, Pablo Campo, and Antonio Gil-Nagel. 2019. "Olfactory function in focal epilepsies: Understanding mesial temporal lobe epilepsy beyond the hippocampus." *Epilepsia open* no. 4 (3):487-492.

Ey, E. H., B. K. Han, R. B. Towbin, and W. K. Jaun. 1988. "Bony inlet stenosis as a cause of nasal airway obstruction." *Radiology* no. 168 (2):477-9.

Favre, J. J., P. Chaffanjon, J. G. Passagia, and J. P. Chirossel. 1995. "Blood supply of the olfactory nerve. Meningeal relationships and surgical relevance." *Surg Radiol Anat* no. 17 (2):133-8, 12-4.

Grossmann, N., I. Brin, D. Aizenbud, J. Y. Sichel, R. Gross-Isseroff, and J. Steiner. 2005. "Nasal airflow and olfactory function after the repair of cleft palate (with and without cleft lip)." *Oral Surg Oral Med Oral Pathol Oral Radiol Endod* no. 100 (5):539-44.

Guilmin-Crépon, Sophie, Catherine Garel, Clarisse Baumann, Dominique Brémond-Gignac, Isabelle Bailleul-Forestier, Suzel Magnier, Mireille Castanet, Paul Czernichow, Thierry Van Den Abbeele, and Juliane Léger. 2006. "High Proportion of Pituitary Abnormalities and Other Congenital Defects in Children with Congenital Nasal Pyriform Aperture Stenosis." *Pediatric Research* no. 60 (4):478-484.

Habu, M., M. Niiro, M. Toyoshima, Y. Kawano, S. Matsune, and K. Arita. 2009. "Transethmoidal meningoencephalocele involving the olfactory bulb with enlarged foramina of the lamina cribrosa--case report." *Neurol Med Chir (Tokyo)* no. 49 (6):269-72.

Hardelin, J. P., and C. Dodé. 2008. "The complex genetics of Kallmann syndrome: KAL1, FGFR1, FGF8, PROKR2, PROK2, et al." *Sex Dev* no. 2 (4-5):181-93.

Hendrix, Philipp, Christoph J. Griessenauer, Paul Foreman, Mohammadali M. Shoja, Marios Loukas, and R. Shane Tubbs. 2014. "Arterial supply of the upper cranial nerves: A comprehensive review." *Clinical Anatomy* no. 27 (8):1159-1166.

Horie, Nobutaka, Minoru Morikawa, Shuji Fukuda, Kentaro Hayashi, Kazuhiko Suyama, and Izumi Nagata. 2012. "*New variant of persistent primitive olfactory artery associated with a ruptured aneurysm.*" no. 117 (1):26.

Hummel, T., A. Haehner, C. Hummel, I. Croy, and E. Iannilli. 2013. "Lateralized differences in olfactory bulb volume relate to lateralized differences in olfactory function." *Neuroscience* no. 237:51-55.

Hummel, T., B. N. Landis, and K. B. Hüttenbrink. 2011. "Smell and taste disorders." *GMS Curr Top Otorhinolaryngol Head Neck Surg* no. 10:Doc04.

Hummel, Thomas, Sophia Henkel, Simona Negoias, José R. B. Galván, Vasyl Bogdanov, Peter Hopp, Susanne Hallmeyer-Elgner, Johannes Gerber, Ulrike Reuner, and Antje Haehner. 2013. "Olfactory bulb volume in patients with temporal lobe epilepsy." *Journal of Neurology* no. 260 (4):1004-1008.

Inoue, Akihiro, Masahiko Tagawa, Yoshiaki Kumon, Hideaki Watanabe, Daisuke Shoda, Kenji Sugiu, and Takanori Ohnishi. 2014. "Ethmoidal dural arteriovenous fistula with unusual drainage route treated by transarterial embolization." *BMJ case reports* no. 2014:bcr2013 011098.

Jacob, Tony George, and J. M. Kaul. 2014. "Morphology of the olfactory fossa – A new look." *Journal of the Anatomical Society of India* no. 63 (1):30-35.

Jang, Woo-Youl, Shin Jung, Tae-Young Jung, Kyung-Sub Moon, and In-Young Kim. 2013. "Preservation of olfaction in surgery of olfactory groove meningiomas." *Clinical neurology and neurosurgery* no. 115 (8):1288-1292.

Jimbo, Hiroyuki, Yukio Ikeda, Hitoshi Izawa, Kuninori Otsuka, and Jo Haraoka. 2010. *Mixed Pial-Dural Arteriovenous Malformation in the Anterior Cranial Fossa*
—Two Case Reports—." *Neurologia medico-chirurgica* no. 50 (6):470-475.

Keller, J. L., and A. Kacker. 2000. "Choanal atresia, CHARGE association, and congenital nasal stenosis." *Otolaryngol Clin North Am* no. 33 (6):1343-51, viii.

Keros, P. 1962. "[On the practical value of differences in the level of the lamina cribrosa of the ethmoid]." *Z Laryngol Rhinol Otol* no. 41:809-13.

Knorr, J. R., R. L. Ragland, R. S. Brown, and N. Gelber. 1993. "Kallmann syndrome: MR findings." *AJNR Am J Neuroradiol* no. 14 (4):845-51.

Kwong, K. M. 2015. "Current Updates on Choanal Atresia." *Front Pediatr* no. 3:52.

Leboucq, N., N. Menjot de Champfleur, S. Menjot de Champfleur, and A. Bonafé. 2013. "The olfactory system." *Diagnostic and Interventional Imaging* no. 94 (10):985-991.

Leopold, Donald A., Todd A. Loehrl, and James E. Schwob. 2002. "Long-term Follow-up of Surgically Treated Phantosmia." *Archives of Otolaryngology–Head & Neck Surgery* no. 128 (6):642-647.

Lesciotto, Kate M., Yann Heuzé, Ethylin Wang Jabs, Joseph M. Bernstein, and Joan T. Richtsmeier. 2018. "Choanal Atresia and Craniosynostosis: Development and Disease." *Plastic and reconstructive surgery* no. 141 (1):156-168.

Levy, Lucien M., Andrew J. Degnan, Salil Sharma, Linda Kelahan, and Robert I. Henkin. 2012. "Morphological Changes of Olfactory Bulbs and Grooves: Initial Report of Supernumerary Olfactory Bulbs." *Journal of Computer Assisted Tomography* no. 36 (4).

Li, Jia, Cheng-zhi Gu, Jian-bin Su, Lian-hai Zhu, Yong Zhou, Huai-yu Huang, and Chun-feng Liu. 2016. "Changes in Olfactory Bulb Volume in Parkinson's Disease: A Systematic Review and Meta-Analysis." *PLOS ONE* no. 11 (2):e0149286.

López-Elizalde, R., A. Campero, T. Sánchez-Delgadillo, Y. Lemus-Rodríguez, M. I. López-González, and M. Godínez-Rubí. 2018. "Anatomy of the olfactory nerve: A comprehensive review with cadaveric dissection." *Clinical Anatomy* no. 31 (1):109-117.

Losee, J. E., R. E. Kirschner, L. A. Whitaker, and S. P. Bartlett. 2004. "Congenital nasal anomalies: a classification scheme." *Plast Reconstr Surg* no. 113 (2):676-89.

Madan, R., V. Sawlani, S. Gupta, and R. V. Phadke. 2004. "MRI findings in Kallmann syndrome." *Neurol India* no. 52 (4):501-3.

Madani, Gisma A., Abdelmoneim S. El-mardi, and Wael Amin Nasr El-Din. 2020. "Analysis of the anatomic variations of the ethmoid roof among Saudipopulation: A radiological study." *Eur. J. Anat.* no. 24: 121-128.

Mak, Y. Erica, Katharine B. Simmons, Darren R. Gitelman, and Dana M. Small. 2005. "Taste and olfactory intensity perception changes following left insular stroke." *Behavioral neuroscience* no. 119 (6):1693-1700.

Massey, E. Wayne, and Bruce Schoenberg. 1984. "Foster Kennedy Syndrome." *Archives of Neurology* no. 41 (6):658-659.

Massimi, Luca, Alessandro Izzo, Giovanna Paternoster, Paolo Frassanito, and Concezio Di Rocco. 2016. "Arachnoid cyst: a further anomaly associated with Kallmann syndrome?" *Child's Nervous System* no. 32 (9):1607-1614.

May, Maureen A., Carla A. Sanchez, Frederic W. B. Deleyiannis, Mary L. Marazita, and Seth M. Weinberg. 2015. "Evidence of olfactory deficits as part of the phenotypic spectrum of nonsyndromic orofacial clefting." *The Journal of craniofacial surgery* no. 26 (1):84-86.

McCabe, Mark J., Carles Gaston-Massuet, Vaitsa Tziaferi, Louise C. Gregory, Kyriaki S. Alatzoglou, Massimo Signore, Eduardo Puelles, Dianne Gerrelli, I. Sadaf Farooqi, Jamal Raza, Joanna Walker, Scott I. Kavanaugh, Pei-San Tsai, Nelly Pitteloud, Juan-Pedro Martinez-Barbera, and Mehul T. Dattani. 2011. "Novel FGF8 Mutations Associated with Recessive Holoprosencephaly, Craniofacial Defects, and Hypothalamo-Pituitary Dysfunction." *The Journal of Clinical Endocrinology & Metabolism* no. 96 (10):E1709-E1718.

McMains, K. Christopher. 2008. "Safety in endoscopic sinus surgery." *Current opinion in otolaryngology & head and neck surgery* no. 16 (3):247-251.

Mondal, Uttam, and Rameshwar Prasad. 2016. "Congenital Arhinia: A Rare Case Report and Review of Literature." *Indian journal of otolaryngology and head and neck surgery: Official publication of the Association of Otolaryngologists of India* no. 68 (4):537-539.

Monkhouse, Stanley. 2005. *Cranial Nerves: Functional Anatomy*. Cambridge: Cambridge University Press.

Morton, W. R. M. 1947. "Arhinencephaly and multiple developmental anomalies occurring in a human full-term foetus." *The Anatomical Record* no. 98 (1):45-58.

Nagayama, Shin, Ryota Homma, and Fumiaki Imamura. 2014. "Neuronal organization of olfactory bulb circuits." *Frontiers in Neural Circuits* no. 8 (98).

Negoias, S., I. Croy, J. Gerber, S. Puschmann, K. Petrowski, P. Joraschky, and T. Hummel. 2010. "Reduced olfactory bulb volume and olfactory sensitivity in patients with acute major depression." *Neuroscience* no. 169 (1):415-421.

Negoias, S., T. Hummel, A. Symmank, J. Schellong, P. Joraschky, and I. Croy. 2016. "Olfactory bulb volume predicts therapeutic outcome in major depression disorder." *Brain Imaging Behav* no. 10 (2):367-72.

Olsen Ø, E., K. Gjelland, H. Reigstad, and K. Rosendahl. 2001. "Congenital absence of the nose: a case report and literature review." *Pediatr Radiol* no. 31 (4):225-32.

Paetau, A., R. Salonen, and M. Haltia. 1985. "Brain pathology in the Meckel syndrome: a study of 59 cases." *Clin Neuropathol* no. 4 (2):56-62.

Panigrahy, Ashok, Vincent Lee, Rafael Ceschin, Giulio Zuccoli, Nancy Beluk, Omar Khalifa, Jodie K. Votava-Smith, Mark DeBrunner, Ricardo Munoz, Yuliya Domnina, Victor Morell, Peter Wearden, Joan Sanchez De Toledo, William Devine, Maliha Zahid, and Cecilia W. Lo. 2016. "Brain Dysplasia Associated with Ciliary Dysfunction in Infants with Congenital Heart Disease." *The Journal of Pediatrics* no. 178:141-148.e1.

Petty, P. G. 1965. "Absence of olfactory bulbs and tracts with an associated bilobed pineal body." *Med J Aust* no. 2 (12):492-4.

Pingault, Veronique, Virginie Bodereau, Viviane Baral, Severine Marcos, Yuli Watanabe, Asma Chaoui, Corinne Fouveaut, Chrystel Leroy, Odile Vérier-Mine, Christine Francannet, Delphine Dupin-Deguine, Françoise Archambeaud, François-Joseph Kurtz, Jacques Young, Jérôme Bertherat, Sandrine Marlin, Michel Goossens, Jean-Pierre Hardelin, Catherine Dodé, and Nadege Bondurand. 2013. "Loss-of-Function Mutations in SOX10 Cause Kallmann Syndrome with Deafness." *The American Journal of Human Genetics* no. 92 (5):707-724.

Pinto, Jayant M. 2011. "Olfaction." *Proceedings of the American Thoracic Society* no. 8 (1):46-52.

Podlesek, Dino, Amir Zolal, Matthias Kirsch, Gabriele Schackert, Thomas Pinzer, and Thomas Hummel. 2019. "Olfactory bulb volume changes associated with trans-sphenoidal pituitary surgery." *PLOS ONE* no. 14 (12):e0224594.

Rai, Rabjot, Joe Iwanaga, Marios Loukas, Rod J. Oskouian, and R. Shane Tubbs. 2018. "Brain Herniation Through the Cribriform Plate: Review and Comparison to Encephaloceles in the Same Region." *Cureus* no. 10 (7):e2961-e2961.

Richman, R. A., P. R. Sheehe, T. McCanty, M. Vespasiano, E. M. Post, S. Guzi, and H. Wright. 1988. "Olfactory deficits in boys with cleft palate." *Pediatrics* no. 82 (6):840-4.

Rienzo, Lino Di, Alberto Artuso, and Cesare Colosimo. 2002. "Isolated Congenital Agenesis of the Olfactory Bulbs and Tracts in a Child without Kallmann's Syndrome." *Annals of Otology, Rhinology & Laryngology* no. 111 (7):657-660.

Rombaux, Philippe, Caroline Huart, Anne G. De Volder, Isabel Cuevas, Laurent Renier, Thierry Duprez, and Cecile Grandin. 2010. "Increased olfactory bulb volume and olfactory function in early blind subjects." *NeuroReport* no. 21 (17).

Rombaux, Philippe, André Mouraux, Bernard Bertrand, Georges Nicolas, Thierry Duprez, and Thomas Hummel. 2006. "Olfactory Function and Olfactory Bulb Volume in Patients with Postinfectious Olfactory Loss." *The Laryngoscope* no. 116 (3):436-439.

Roy, E. P., 3rd, J. L. Frost, and S. S. Schochet, Jr. 1987. "Persistent olfactory bulb ventricle." *Clin Neuropathol* no. 6 (2):86-7.

Sarafoleanu, C., C. Mella, M. Georgescu, and C. Perederco. 2009. "The importance of the olfactory sense in the human behavior and evolution." *Journal of medicine and life* no. 2 (2):196-198.

Sato, Yuiko, Hiroshi Kashimura, Masaru Takeda, Kohei Chida, Yoshitaka Kubo, and Kuniaki Ogasawara. 2015. "Aneurysm of the A1 Segment of the Anterior Cerebral Artery Associated with the Persistent Primitive Olfactory Artery." *World Neurosurgery* no. 84 (6):2079.e7-2079.e9.

Scalia, Frank, and Sarah S. Winans. 1975. "The differential projections of the olfactory bulb and accessory olfactory bulb in mammals." *Journal of Comparative Neurology* no. 161 (1):31-55.

Schwartz, M. L., and L. Savetsky. 1986. "Choanal atresia: clinical features, surgical approach, and long-term follow-up." *Laryngoscope* no. 96 (12):1335-9.

Shemshadi, Hashem, Mojtaba Azimian, Mohammad Ali Onsori, and Mahdi AzizAbadi Farahani. 2008. "Olfactory function following open rhinoplasty: A 6-month follow-up study." *BMC Ear, Nose and Throat Disorders* no. 8 (1):6.

Smitka, M., N. Abolmaali, M. Witt, J. C. Gerber, W. Neuhuber, D. Buschhueter, S. Puschmann, and T. Hummel. 2009. "Olfactory bulb ventricles as a frequent finding in magnetic resonance imaging studies of the olfactory system." *Neuroscience* no. 162 (2):482-485.

Sobin, C., K. Kiley-Brabeck, K. Dale, S. H. Monk, J. Khuri, and M. Karayiorgou. 2006. "Olfactory disorder in children with 22q11 deletion syndrome." *Pediatrics* no. 118 (3):e697-703.

Stewart, R. M. 1939. "Arhinencephaly." *J Neurol Psychiatry* no. 2 (4):303-12.

Suzuki, Jiro, Kazuo Mizoi, and Takashi Yoshimoto. 1986. "*Bifrontal interhemispheric approach to aneurysms of the anterior communicating artery.*" no. 64 (2):183.

Takahashi, M. P., I. Miyai, T. Matsumura, S. Nozaki, and J. Kang. 1997. "[A case of Kallmann syndrome with empty sella and arachnoid cyst]." *Rinsho Shinkeigaku* no. 37 (8):704-7.

Takahashi, Tsutomu, Yumiko Nakamura, Kazue Nakamura, Eiji Ikeda, Atsushi Furuichi, Mikio Kido, Yasuhiro Kawasaki, Kyo Noguchi, Hikaru Seto, and Michio Suzuki. 2013. "Altered depth of the olfactory sulcus in first-episode schizophrenia." *Progress in Neuro-Psychopharmacology and Biological Psychiatry* no. 40:167-172.

Takahashi, Tsutomu, Stephen J. Wood, Alison R. Yung, Barnaby Nelson, Ashleigh Lin, Murat Yücel, Lisa J. Phillips, Yumiko Nakamura, Michio Suzuki, Warrick J. Brewer, Tina M. Proffitt, Patrick D. McGorry, Dennis Velakoulis, and Christos Pantelis. 2014. "Altered depth of the olfactory sulcus in ultra high-risk individuals and patients with psychotic disorders." *Schizophrenia Research* no. 153 (1):18-24.

Tanik, Nermin, Halil Ibrahim Serin, Asuman Celikbilek, Levent Ertugrul Inan, and Fatma Gundogdu. 2015. "Olfactory bulb and olfactory sulcus depths are associated with disease duration and attack frequency in multiple sclerosis patients." *Journal of the Neurological Sciences* no. 358 (1):304-307.

Tavin, E., E. Stecker, and R. Marion. 1994. "Nasal pyriform aperture stenosis and the holoprosencephaly spectrum." *Int J Pediatr Otorhinolaryngol* no. 28 (2-3):199-204.

Tellier, A. L., V. Cormier-Daire, V. Abadie, J. Amiel, S. Sigaudy, D. Bonnet, P. de Lonlay-Debeney, M. P. Morrisseau-Durand, P. Hubert, J. L. Michel, D. Jan, H. Dollfus, C. Baumann, P. Labrune, D. Lacombe, N. Philip, M. LeMerrer, M. L. Briard, A. Munnich, and S. Lyonnet. 1998. "CHARGE syndrome: Report of 47 cases and review." *American Journal of Medical Genetics* no. 76 (5):402-409.

Thompson, Lester D. R. 2009. "Olfactory neuroblastoma." *Head and neck pathology* no. 3 (3):252-259.

Truwit, C. L., A. J. Barkovich, M. M. Grumbach, and J. J. Martini. 1993. "MR imaging of Kallmann syndrome, a genetic disorder of neuronal migration affecting the olfactory and genital systems." *AJNR Am J Neuroradiol* no. 14 (4):827-38.

Tsuji, Takehisa, Masamitsu Abe, and Kazuo Tabuchi. 1995. *Aneurysm of a persistent primitive olfactory artery*. no. 83 (1):138.

Tsutsumi, Satoshi, Hideo Ono, and Yukimasa Yasumoto. 2017. "Visualization of the olfactory nerve using constructive interference in steady state magnetic resonance imaging." *Surgical and Radiologic Anatomy* no. 39 (3):315-321.

Turetsky, Bruce I., Patrick Crutchley, Jeffrey Walker, Raquel E. Gur, and Paul J. Moberg. 2009. "Depth of the olfactory sulcus: A marker of early embryonic disruption in schizophrenia?" *Schizophrenia Research* no. 115 (1):8-11.

Uchino, Akira, Naoko Saito, Eito Kozawa, Waka Mizukoshi, and Kaiji Inoue. 2011. "Persistent primitive olfactory artery: MR angiographic diagnosis." *Surgical and Radiologic Anatomy* no. 33 (3):197-201.

Ulualp, S. O. 2008. "Complications of endoscopic sinus surgery: appropriate management of complications." *Curr Opin Otolaryngol Head Neck Surg* no. 16 (3):252-9.

Vaid, S., and N. Vaid. 2015. "Normal Anatomy and Anatomic Variants of the Paranasal Sinuses on Computed Tomography." *Neuroimaging Clin N Am* no. 25 (4):527-48.

Van Hove, Johan L. K., Gail A. Spiridigliozzi, Ralph Heinz, Allyn McConkie-Rosell, A. Kimberly Iafolla, and Stephen G. Kahler. 1995. "Fryns syndrome survivors and neurologic outcome." *American Journal of Medical Genetics* no. 59 (3):334-340.

Vasović, Ljiljana, Milena Trandafilović, Slobodan Vlajković, Ivan Jovanović, and Slađana Ugrenović. 2013. "Persistent Primitive Olfactory Artery in Serbian Population." *BioMed Research International* no. 2013:903460.

Vaughan, David N., and Graeme D. Jackson. 2014. "The Piriform Cortex and Human Focal Epilepsy." *Frontiers in Neurology* no. 5 (259).

Velho, Vernon, Harish Naik, Pravin Survashe, Sachin Guthe, Anuj Bhide, Laxmikant Bhople, and Amrita Guha. 2019. "Management strategies of cranial encephaloceles: A neurosurgical challenge." *Asian Journal of Neurosurgery* no. 14 (3):718-724.

Wang, Shou-Sen, He-Ping Zheng, Xiang Zhang, Fa-Hui Zhang, Jun-Jie Jing, and Ru-Mi Wang. 2008. "Microanatomy and surgical relevance of the olfactory cistern." *Microsurgery* no. 28 (1):65-70.

Weiss, Tali, Timna Soroka, Lior Gorodisky, Sagit Shushan, Kobi Snitz, Reut Weissgross, Edna Furman-Haran, Thijs Dhollander, and Noam Sobel. 2020. "Human Olfaction without Apparent Olfactory Bulbs." *Neuron* no. 105 (1):35-45.e5.

Yaldizli, Ö., I.-K. Penner, T. Yonekawa, Y. Naegelin, J. Kuhle, M. Pardini, D. T. Chard, C. Stippich, J.-i. Kira, K. Bendfeldt, M. Amann, E.-W. Radue, L. Kappos, and T. Sprenger. 2016. "The association between olfactory bulb volume, cognitive dysfunction, physical disability and depression in multiple sclerosis." *European Journal of Neurology* no. 23 (3):510-519.

Yamamoto, Tetsuya, Kensuke Suzuki, Tomosato Yamazaki, Wataro Tsuruta, Takao Tsurubuchi, and Akira Matsumura. 2009. "Persistent Primitive Olfactory Artery Aneurysm." *Neurologia medico-chirurgica* no. 49 (7):303-305.

Yousem, D. M., R. J. Geckle, W. Bilker, D. A. McKeown, and R. L. Doty. 1996. "MR evaluation of patients with congenital hyposmia or anosmia." *American Journal of Roentgenology* no. 166 (2):439-443.

Yousem, D. M., W. J. Turner, C. Li, P. J. Snyder, and R. L. Doty. 1993. "Kallmann syndrome: MR evaluation of olfactory system." *AJNR Am J Neuroradiol* no. 14 (4):839-43.

Zhan, Xiaojun, Yongxiang Wei, Xutao Miao, Cong Zhang, and Demin Han. 2007. "[Effects of ischemia on olfactory bulb in rats]." *Lin chuang er bi yan hou tou jing wai ke za zhi = Journal of clinical otorhinolaryngology, head, and neck surgery* no. 21 (5):219-221.

In: Cranial Nerves
Editor: Thomas M. Yi

ISBN: 978-1-53618-823-3
© 2021 Nova Science Publishers, Inc.

Chapter 2

CRANIAL NERVE II: OPTIC NERVE; ANATOMY, EVALUATION, PATHOLOGY AND SURGICAL APPROACHES

Natalie Homer[1], Aliza Epstein[2] and Craig Kemper[3]

[1]Department of Ophthalmology, University of California at Davis
[2]Department of Oculoplastic Surgery,
TOC Eye and Face, Austin, TX US
[3]Affiliate Faculty, Dept. of Neurosurgery, Dell Medical School,
The University of Texas at Austin

ABSTRACT

The second cranial nerve is the optic nerve and is a vital component of the visual pathway. The optic nerve spans from the eye to the optic chiasm where the two optic nerves fuse. The optic nerve can be divided into its various segments including the intraocular, intraorbital, intracanalicular, and intracranial portions of the nerve. Each segment has a unique blood supply based on location.

The structure and function of the optic nerve can be analyzed by a number of means. The optic nerve head can be examined directly by ophthalmoscopic exam or can be visualized by a variety of imaging modalities including optical coherence tomography (OCT), confocal scanning laser ophthalmoscopy (CSLO), and scanning laser polarimetry (SLP). Neuro-imaging with computed tomography (CT) and magnetic resonance imaging (MRI) can also provide visualization of the orbit and optic nerve. A variety of tests can provide information on the functioning of the optic nerve. These include visual acuity testing, the pupillary exam, color vision assessment, and visual field testing.

PART 1: OPTIC NERVE ANATOMY AND FUNCTION

Introduction and Embryology

The optic nerve is the second cranial nerve, responsible for transmitting afferent visual input from the retina to the brain. Studies have suggested the presence of efferent nerve fibers as well, though their role has not been fully elucidated [1]. In ocular development, the optic nerve begins as the optic stalk, which connects the optic vesicle to the developing forebrain [2]. The stalk is comprised of neuroectodermal cells surrounded by neural crest cells. At six weeks gestation, the inner retinal ganglion cell fibers migrate to fill the stalk lumen, forming the optic nerve. Subsequent maturation of the neuroectodermal cells and neural crest cells into oligodendrocytes and meninges, respectively, completes nerve development.

Segments of the Optic Nerve

The optic nerve contains approximately 1.0-1.2 million retinal ganglion cell axons. The cell bodies of these axons comprise the ganglion cell layer of the retina and the axons travel within the retinal nerve fiber layer (RNFL) and coalesce to form the optic nerve. The optic nerve can be divided into four different segments: intraocular, intraorbital, intracanalicular, and intracranial.

The intraocular portion of the optic nerve is 1 mm in length and can be visualized during an ophthalmoscopic exam as the optic nerve head or optic disc. The optic disc is located 3 mm nasal and 0.8 mm superior to the foveola (center) of the macula [2]. The average size of the optic disc is 1.76 mm horizontally and 1.92 mm vertically. The central retinal artery and vein pass through the center of the optic disc in an axon free region known as the optic cup. Atrophy of optic nerve fibers manifests as pathologic enlargement of the optic nerve cup. All layers of the retina, except for the retinal ganglion cell axons, end at the optic disc, correlating with the blind spot seen on visual field testing. The nonmyelinated retinal ganglion axons maintain retinotopic organization as they coalesce to form the optic nerve. Starting at the optic nerve head, astrocytic glial cells divide the axons into bundles, or fascicles, through pores in the lamina cribrosa at the level of the scleral foramen.

The lamina cribrosa is comprised of connective tissue including collagen (type I and III), elastin, laminin, and fibronectin. As the nerve bundles exit the lamina cribrosa posteriorly, the diameter of the optic nerve expands to 3 mm due to myelination of the nerve fibers by oligodendrocytes. The intraocular portion of the optic nerve can be further divided into the prelaminar, laminar, and retrolaminar areas based on relative location to the lamina cribrosa.

The intraorbital portion of the optic nerve is approximately 25-30 mm in length and lies within the muscle cone of the orbit. The distance from the posterior globe to the optic foramen is 18 mm. This redundancy in optic nerve length allows for rotational and axial movement of the globe without harmful optic nerve traction. The meningeal sheath layers, including the dura, arachnoid, and pia, envelop the optic nerve as it exits the globe allowing circulation of cerebrospinal fluid around the optic nerve. The central retinal artery and vein enter the optic nerve 12 mm posterior to the globe. Before passing through the optic foramen, the optic nerve passes through the annulus of Zinn, a circular structure made of the origins of the four extraocular rectus muscles. At the optic foramen, the dural sheath fuses with the periosteum, firmly fixating the optic nerve into position. The arachnoid sheath that surrounds the intraorbital optic nerve is continuous with the arachnoid of the subdural intracranial space.

The portion of the optic nerve within the optic canal is termed the intracanalicular segment. The optic canal runs superiorly and medially through the lesser wing of the sphenoid bone, measuring approximately 8-10 mm in length and 5-7 mm in width. The sphenoid sinus lies medial to the canal. Within the canal, the optic nerve is firmly attached to the periosteum making it susceptible to traumatic injury by shearing forces. The ophthalmic artery and sympathetic nerves accompany the optic nerve within the canal.

The intracranial portion of the optic nerve is 8-12 mm long and is the final portion of the optic nerve. As the optic nerve becomes intracranial it passes under a fold of dura mater called the falciform ligament. The anterior loop of the internal carotid artery lies just inferior and lateral to the optic nerve. The proximal anterior cerebral arteries pass over the optic nerve, and are connected by the anterior communicating artery to form the anterior portion of the circle of Willis. The optic nerves pass posteriorly over the cavernous sinus to join and form the optic chiasm. The optic nerve lacks a sheath in its intracranial portion.

The optic nerves fuse to form the optic chiasm, the point at which nasal axons cross to the contralateral side. These crossing axons comprise approximately 53% of total nerve axons. Temporal axons do not decussate at the optic chiasm. The axons of the optic nerve enter the brain and terminate in four nuclei: the *lateral geniculate nucleus* of the thalamus, the *superior colliculus* of the midbrain, the *pretectum* of the midbrain and the suprachiasmatic nucleus of the *hypothalamus [4]*. Due to nasal fiber decussation at the optic chiasm, visual information from the left visual field is transmitted to the right optic tract and visual information from the right visual field is transmitted to the left optic tract. Optic radiations originate from the lateral geniculate nucleus and transmit visual information to the primary visual cortex. The superior colliculus of the midbrain plays a role in coordinating eye movements as they relate to head and body position. The pretectal nucleus of the midbrain is involved in the pupillary light reflex. The suprachiasmatic nucleus of the hypothalamus is part of the retinohypothalamic pathway that is involved in circadian rhythm development.

Blood Supply to the Optic Nerve

The blood supply to the optic nerve varies along the course of the nerve and involves contributions from multiple sources including the pial vessels, the central retinal artery, and other branches of the ophthalmic artery [5]. The prelaminar region of the optic nerve is supplied by posterior ciliary arteries and recurrent choroidal arteries. The laminar portion of the optic nerve is supplied by short posterior ciliary arteries directly or from branches of the circle of Zinn-Haller, which is formed by a circular anastomoses of short posterior ciliary arteries embedded in the sclera at the level of the lamina cribrosa. A pial vascular plexus supplied by pial branches from the circle of Zinn Haller and short posterior ciliary arteries supplies the retrolaminar region. Additionally, pial branches of the central retinal artery can supply this retrolaminar region.

The intraorbital portion of the optic nerve is supplied by pial vessels and branches of the ophthalmic artery. The distal portion of the intraorbital optic nerve is also supplied by central retinal artery intraneural branches. The ophthalmic artery supplies the intracanalicular optic nerve. The intracranial optic nerve is supplied by branches of the ophthalmic artery and the internal carotid artery.

PART 2: EXAMINATION OF THE OPTIC NERVE

The optic nerve can be examined both anatomically and functionally. Direct visualization can be performed by clinical examination. The direct ophthalmoscope can provide an upright and monocular view of the optic nerve with 15-times magnification. It is important for the light aperture of the direct ophthalmoscope to match the diameter of the pupil in order to avoid glare and constriction of the pupil. An indirect ophthalmoscopic exam provides another means for examination of the posterior segment, however due to reduced magnification, this is better suited for evaluation of the retina than the optic nerve. Slit lamp examination of the optic nerve with the use of hand-held lenses is an ideal method for visualization as it provides an inverted and stereoscopic view of the nerve that allows for precise discrimination of fine details of the optic nerve disc, cup, and vessels. Changes in the optic nerve can be examined over time by use of optic nerve stereo photography.

The Optic nerve maintains a discreet arrangement of axons within its course and projects this functional architecture through the visual pathway. Clinically diagnosis can accurately be made as to disruption or interference along the optic pathway by demonstrating the pattern of visual loss.

Imaging of the Optic Nerve

There are a variety of measures to provide optic nerve imaging. Optical coherence tomography (OCT), confocal scanning laser ophthalmoscopy (CSLO), and scanning laser olarimetry (SLP) are imaging technologies that provide microscopic assessment of the axonal fibers of the RNFL and coalesce to form the optic nerve. OCT is the modality most commonly utilized by ophthalmologists to provide objective assessment of the optic nerve. Thickness of the RNFL is assessed in the 4 quadrants of the optic nerve head, with thinning or thickening (e.g., in the setting of edema) suggesting nerve pathology. Glaucomatous optic nerve damage classically induces retinal nerve fiber atrophy in the decreasing order of inferior, superior, nasal, temporal quadrants, referred to as the "ISNT" rule. Normative databases are overlaid with these various technologies to distinguish optic nerve damage from normal variation. Progression of RNFL thinning can be objectively surveilled in glaucoma or other optic neuropathies with use of OCT.

Computed tomography (CT) and magnetic resonance imaging (MRI) are two commonly used neuroimaging modalities for imaging of the orbit and optic nerve. Advantages of CT scanning include low cost, rapid image acquisition, and excellent spatial resolution, with the disadvantages of exposure to ionizing radiation and poor resolution of the orbital apex. MRI offers advantages over CT with better resolution of the optic nerve and orbital apex at a greater cost compared to CT [7]. MRI imaging should include short tau inversion recovery (STIR) and T2 fat-saturated sequences of the orbit as well as diffusion-weighted imaging (DWI), fluid-attenuated inversion recovery (FLAIR), and susceptibility-weighted imaging (SWI) for the entire brain, when indicated. Imaging may reveal an intraorbital tumor involving the optic nerve as in an optic nerve meningioma, or enhancement of the nerve suggestive of inflammation or optic neuritis.

Functional Tests of the Optic Nerve

Visual acuity assessment provides a nonspecific measure of optic nerve function. Decreased visual acuity may be due to ocular, optic nerve, or intracranial pathology, and therefore a thorough eye exam must be performed to exclude alternative causes for vision loss. Examples of such ocular pathology which spares the optic nerve but leads to decreased vision includes corneal ulcer, cataract, and retinal detachment.

The pupillary exam is essential for assessing function of the optic nerves, relying on the pupillary light reflex afferent and efferent pathways. In the afferent pathway, sensory input (i.e., detection of light) from the retina is transmitted to the pretectal nucleus of the midbrain by way of the optic nerve, optic chiasm, optic tract, lateral geniculate nucleus. From the pretectal nucleus, fibers distribute to the Edinger-Westphal nuclei of the oculomotor complex bilaterally. Efferent bilateral preganglionic parasympathetic fibers travel on the oculomotor nerve, synapse in the ciliary ganglion, and prompt postganglionic parasympathetic fibers within the short ciliary nerves of the iris sphincter muscle to induce pupillary constriction. Nerve fibers decussate at two points along this pathway: the optic chiasm and the pretectal nuclei. Shining light into one eye induces bilateral pupillary constriction due to a direct constriction of pupil receiving light and consensual constriction of pupil not receiving light.

Lack of pupillary constriction to a light stimulus suggests pathology of either the afferent or efferent pupillary pathway.

The swinging light test can be performed to identify a relative afferent pupillary defect (rAPD) [6]. In absence of pathology, quickly alternating a light between the two eyes should not elicit a change in pupil size due to sustained direct and consensual pupillary responses. In the setting of unilateral optic nerve damage, the degree of pupillary constriction observed in the direct response in the affected eye and consensual response in the unaffected eye is less than the pupillary constriction observed by direct response in the unaffected eye. When the light moves quickly from the affected to unaffected eye, the optic nerve in the unaffected eye will constrict due to a relative increase in afferent visual input. When the light moves back to the affected eye, the pupil of the affected eye will dilate due a relative decrease in afferent visual input. Contralateral rAPD may occur in the setting of an isolated optic tract lesion as there are more crossed than uncrossed fibers at the optic chiasm. While an afferent pupillary defect is most suggestive of optic nerve abnormality, occasionally retinal pathology can induce a subtle rAPD if there is significant and asymmetric damage, as in a central retinal artery occlusion, central retinal vein occlusion, or age related macular degeneration. Neutral density filters, cross-polarized filters, and subjective grading can be used to quantify the severity of the rAPD. The density of the filter required to neutralize the APD defines the severity of the APD.

Color vision assessment is essential in evaluation of optic nerve function. Decreased color vision, or dyschromatopsia, can be found in congenital or acquired optic nerve or macular pathology, the latter involving damage of the photoreceptor cells of the retina. Intact visual acuity paired with a color vision deficit is more suggestive of optic nerve pathology, whereas retinal pathology may exhibit a similar degree of decreased visual acuity and color vision. Persistent dyschromatopsia after recovery of visual acuity is suggestive of an optic neuropathy. Abnormal color vision in the affected eye paired with normal color vision in the unaffected eye suggests acquired pathology rather than congenital or hereditary color blindness. Color vision can be tested by a variety of methods including American Optical Hardy-Rand-Rittler and Ishhara color plates, or color arrangement tests. Color plates can provide a quick and gross assessment of color vision, by asking the patient to identify a symbol comprised of multicolored dots embedded in a background of a different color. Arrangement tests, such as the Farnsworth Panel D-15 test and the Farnsworth-Munsell 100-hue test, requiring patients to arrange colored discs based on hue and intensity, provide a more sensitive assessment of color vision. Lastly, a red desaturation test can serve as a quick screening tool for color vision abnormality and optic nerve function by asking a patient to rate the intensity of a red object as seen by one eye compared to the other. Patients with optic neuropathy will commonly report decreased intensity of the red color in the affected eye. Color vision testing should always be performed under monocular conditions.

Visual field testing provides a subjective analysis of optic nerve function as it depends on cooperation of the patient being tested. It is important to test visual fields under monocular conditions. Confrontational visual field testing is a simple and quick method to assess peripheral vision though it lacks sensitivity. The patient is asked to look at a fixed central target (e.g., the examiner's eye or nose) while covering the other eye, while the examiner presents visual cues (e.g., holding up fingers to be counted) in each quadrant of the patient's visual fields. Tangent screen visual field testing is another method that can be utilized with ease in an office setting using a black felt background and circular stitching positioned at five-degree increments.

Automatic static perimetry visual field testing is the standard method for assessing peripheral vision. With this method, threshold sensitivity measurements are presented at a number of test locations while the patient is focused on a central target image [8]. Again this is completed under monocular viewing conditions. The Humphrey Field Analyzer (HFA) perimeter (Carl-Zeiss Meditec, Dublin, CA) is a commonly used system to provide visual field testing for patients. The test can assess various degrees of peripheral visual fields (including 10, 24, or 30 degrees). Manual kinetic perimetry may be performed for patients who cannot perform the automated test as this testing modality involves a skilled examiner physically moving light targets throughout the visual field compared to flashing targets utilized in automated perimetry

Any intraocular pathology that decreases a patient's vision can influence their visual field testing. For example, dense cataracts or macular disease can lead to a visual field defect that is not specific to the optic nerve. However, in the absence of concomitant significant ocular pathology, visual field defect analysis can be used to localize a pathology to a specific portion of the optic nerve pathway and monitor for optic disease progression (e.g., in glaucoma). Multiple visual field results are needed to verify pathology as reliability of the visual field test highly depends on patient cooperation.

PART 3: OPTIC NERVE PATHOLOGY

Intraocular Optic Nerve Pathology

Congenital Optic Neuropathies

Optic neuropathies may be inherited or acquired [9]. The most common hereditary optic neuropathy is dominant optic atrophy (DOA), an autosomally-dominant inherited disorder resulting in premature degeneration of the optic nerve. Patients often present in the first decade of life with decreased vision, dyschromatopsia and central scotomas, and are found clinically to have triangular pallor and excavation of the temporal optic disc. Leber's Hereditary Optic Neuropathy (LHON) is a mitochondrially inherited DNA abnormality that leads to the degeneration of the retinal ganglion cells. This disorder commonly presents with visual acuity decline in the second or third decade of life. Clinically, patients are observed to have optic nerve head hyperemia, with vessel dilation and tortuosity, and occasional macular edema and hemorrhages. While there is currently no proven treatment for these disorders, promising studies of use of quinolone analogs and gene therapies are underway that may someday prove to be curative.

In normal retinal development, myelination terminates at the lamina cribrosa, sparing the retinal nerve fiber layer. Abnormally myelinated retinal nerve fibers have been found in approximately 1% of autopsy studies [10]. This is suspected to occur from continued myelination beyond the lamina cribrosa. Clinically these appear as white-gray straited patches with feathered borders spanning across the retinal surface. These may be asymptomatic or associated with visual field abnormalities and have no effective treatment.

Congenital optic nerve hypoplasia may be seen clinically as an underdeveloped optic nerve head displaying the classic "double ring" sign. Visual acuity can range from normal to significant impairment, often with associated strabismus and nystagmus. Optic nerve hypoplasia may be associated with septo-optic dysplasia, or de Morsier Syndrome, often found to have pituitary hypoplasia and midline cerebral malformations, including absence of the septum pellucidum or thinning of the corpus callosum. Patients with such findings on neuroimaging should be evaluated for concomitant endocrine abnormalities [11]. Optic nerve hypoplasia has also been found to be associated with fetal alcohol syndrome [12] and Down's syndrome [13].

Acquired Optic Neuropathies

The most commonly acquired optic neuropathy, and leading cause of worldwide irreversible vision loss, is glaucoma [14]. Although the pathogenesis of glaucoma is contested, the most widely accepted mechanism relates to elevated intraocular pressure inducing ganglion cell death. An imbalance balance between aqueous humor production by the ciliary body and its drainage through ocular outflow pathways, leads to elevated intraocular pressure, and theoretical pressure-induced nerve atrophy. Treatment is, thus, targeted at decreasing aqueous production or increasing outflow, through topical ophthalmic medication, targeted laser therapy and surgical intervention.

Ischemic Optic Neuropathies

Ischemic optic neuropathies are common causes of acquired vision loss in the adult population. Ischemic optic neuropathy (ION) is classified as anterior or posterior, depending on segment of optic nerve affected, with anterior accounting for 90% of cases [15]. Anterior cases are further divided into non-arteritic and arteritic, that resulting from temporal arteritis. Risk factors for the non-arteritic form include hypertension, diabetes mellitus, hyperlipidemia, stroke, ischemic heart disease, tobacco use, systemic atherosclerosis and obstructive sleep apnea [15]. A small cup-to-disc ratio is also believed to be a risk factor. Patients with temporal arteritis are at risk for arteritic ION, and frequently have coinciding symptoms of temple and scalp tenderness, jaw claudication, proximal musculoskeletal aches and elevated serum inflammatory markers. Diagnosis can be confirmed via temporal artery biopsy and prompt treatment initiated with high-dose systemic steroids.

Toxic Optic Neuropathies

Optic neuropathy may ensue following excessive alcohol or tobacco use, dietary deficiencies in vitamin B12 and thiamine, as well as following toxic exposure to methanol and ethylene glycol. Medications most commonly associated with optic neuropathy include ethambutol, isoniazid, amiodarone, and tamoxifen. Other medications reported to induce optic neuropathy include methotrexate, cyclosporine, vincristine, cisplatin and chloramphenicol [16]. Visual field testing most commonly shows a central or cecocentral defect in the setting of toxic and medication-induced optic neuropathy [16] and management includes cessation of the offending agent.

Radiation-Induced Optic Neuropathy

Patients with history of radiation to the brain or orbit may develop radiation optic neuropathy (RON) due to endothelial cell ischemia [16]. The risk in increased in patients who've also received chemotherapy [17]. Patients may present with an edematous or normal-appearing optic nerve, often with sign of radiation retinopathy with retinal hemorrhages, cotton wool spots, exudates and macular edema. MRI imaging may show optic nerve enhancement. Despite management with steroids, antiplatelet drugs, anticoagulants, and hyperbaric oxygen, with poor visual prognosis [18].

Intraorbital Optic Nerve Pathology

Acute Demyelinating Optic Neuritis

Demyelinative optic neuritis is an acute autoimmune disorder of the optic nerve that typically afflicts young adults. This disease most commonly presents with painful subacute unilateral vision loss with decreased color vision and contrast sensitivity [19]. Periocular discomfort is classically increased with eye movements. The Uhthoff phenomenon, a worsening of vision with increased body temperature from physical exertion or with taking a hot shower, is specific but only present in 50% of those affected [20]. Clinically the optic disc appears normal (non-edematous) in 66-75% of cases due to primarily retrobulbar nerve involvement. Gadolinium-enhanced MRI imaging should be performed to evaluate for central demyelinating lesions and to rule out a compressive mass lesion. MRI imaging shows optic nerve sheath enhancement in up to 75% of cases of optic neuritis [19].

The optic neuritis treatment trial found 50% of affected patients to develop multiple sclerosis (MS) within the subsequent 15 years (21 - ONSG) [21]. This risk strongly correlated with presence or absence of demyelinating white matter lesions on baseline non-contrast-enhanced MRI exam, with 72% risk and 25% risk, respectively. The presence of optic disc swelling portended a lower risk of developing MS.

Management of optic neuritis may be with observation or systemic steroids. The ONTT found that final visual acuity was not significantly altered with use of steroid treatment [22], though time to improvement was reduced. Those patients who received oral steroids had a higher incidence of recurrence during the study period. Therefore, the use of oral steroids as a primary treatment is not advised in this disease. Lack of improvement after 4–6 weeks is atypical of optic neuritis and should prompt evaluation for an alternative diagnosis, including inflammatory, infiltrative or compressive optic neuropathy [23].

Neuromyelitis Optica

Neuromyelitis optica (NMO), or Devic's disease, is defined by optic neuritis accompanied by transverse or ascending myelopathy. This condition typically presents in young patients of Asian or African descent [16]. Unlike MS-associated demyelinative optic neuritis, NMO is often associated with severe and bilateral vision loss. Diagnostic criteria includes an MRI of the spine demonstrating demyelination extending over 3 or more vertebral segments. Serum testing for the aquaporin-4 (AQP4) antibody is also useful in the diagnosis, with 62.8% sensitivity and 98.3% specificity (24 Paul F). Acute management includes high-dose intravenous steroids, followed by immunomodulatory maintenance therapy [25].

Optic Nerve Inflammation

Systemic granulomatous disease processes can afflict the optic nerve via primary or secondary infiltration of the optic chiasm or optic tracts. Sarcoidosis may induce optic nerve inflammation of two subtypes: subacute inflammatory presentation similar to that of optic neuritis, and indolent optic neuropathy. Optic perineuritis and chiasmal inflammation have also been reported. Systemic evaluation for sarcoidosis, including serum studies and pulmonary imaging, can often elucidate the diagnosis. Treatment with corticosteroids is effective in a majority of cases [26].

Granulomatosis with polyangiitis, often originating in the paranasal sinuses, may infiltrate the orbital apex and directly infiltrate the optic nerve, inducing a painless optic neuropathy, as well as ophthalmoplegia. Management often requires sinus surgery as well as systemic anti-inflammatory treatment with high-dose steroids or cyclophosphamide initially, followed by maintenance therapy, often with azathioprine or methotrexate [27].

Idiopathic orbital inflammation may occur in all age groups. These lesions are characterized by lack of systemic or serum findings, non-specific inflammation and brisk responsiveness to steroids.

Thyroid Orbitopathy

Compressive optic neuropathy may ensue in fewer than 5% of patients with thyroid orbitopathy. Orbital muscle and soft tissue enlargement may lead to mechanical compression of the optic nerve within the orbital apex [28]. Others hypothesize that stretch of the optic nerve from progressive proptosis may lead to optic nerve compromise [29]. Management is performed with a combination of thyroid control, systemic anti-inflammatory and immunomodulatory therapy, radiation, and surgical decompression.

Traumatic Optic Neuropathy

Traumatic optic neuropathy (TON) may be seen following blunt or penetrating facial trauma, with indirect or direct optic nerve injury. Blunt trauma to the periorbital area may transmit shock injury to the intracanalicular optic nerve, leading to transient or permanent dysfunction. Direct optic nerve injury may occur from bony fragment impingement within the optic canal or orbit. While high-dose steroids have occasionally been advocated in the treatment of TON, the Corticosteroid Randomization after Significant Head Trauma (CRASH) study showed an increased relative risk of death in the setting of significant head injury [30]. Steroid treatment is therefore not recommended as routine treatment.

Optic nerve compression injury may also be seen in the setting of traumatic orbital emphysema or retrobulbar hematoma. Blunt trauma resulting in an orbital wall fracture may allow air entry into the orbit cavity from the adjacent sinus, with intraorbital optic nerve compression. Similarly, hemorrhage within the enclosed orbital space may induce compression. Elevation in intraocular pressure may indicate impending optic nerve damage. Patients may show other signs of acute orbital syndrome, including ophthalmoplegia and proptosis. Mild cases may be observed, while cases of increased intraocular pressure should be managed by emergent lateral canthotomy and lateral canthal tendon lysis or by needle decompression until the pressure is normalized [31].

Primary Tumors of the Optic Nerve

Optic Nerve Glioma

Optic nerve gliomas are the most common optic pathway tumor found in children. These lesions are astrocytic in origin and typically carry a benign pathology in children (i.e., pilocytic astrocytoma), though may be seen to have more aggressive pathology in adult patients. Gliomas may occur anywhere along the optic pathway, with 25% affecting the intraorbital optic nerve. Patients typically present in children or adolescence with decrease visual acuity or dyschromotopsia, abnormal pupillary function, painless proptosis. Imaging most commonly reveals fusiform enlargement of the optic nerve, often with kinking and/or tortuosity, without calcification.

Optic nerve gliomas may be sporadic or associated with neurofibromatosis type 1, with sporadic lesions found to portend a worse diagnosis [32]. The presence of bilateral optic nerve gliomas is virtually diagnostic for neurofibromatosis (NF) [33]. There is increasing evidence that optic nerve gliomas associated with NF1 behave differently than those obtained sporadically [34]. NF1-associated lesions tend to be intraorbital, with chance of spontaneous regression [14], whereas sporadic lesions are more commonly involving the chiasm and have a higher rate of visual loss (80% vs. 20%) [33]. Imaging features of non-NF cases most commonly includes fusiform nerve enlargement with an intact dural sheath, whereas NF-associated lesions more often display optic nerve thickening and kinking [35, 36].

It may be difficult to accurately assess on imaging the extent of an optic pathway glioma. The optic foramen may be distended in lesions restricted to the intraorbital segment of the optic nerve due to secondary meningeal hyperplasia [37]. Cases of a normal-appearing optic foramen on imaging have been later confirmed surgically to intercanalicular disease extension. Thus clinical surveillance, rather than radiologic, may be more sensitive [38, 39].

Histopathologically these lesions are classically found to have glial cell proliferation within enlarged nerve bundles, spindle-shaped pilocytic (hair-like) astrocytes, and eosinophilic Rosenthal fibers [37].

Optic nerve gliomas presenting in adulthood carry increased chance for malignant pathology. These tumors may present with monocular blurring or ipsilateral retrobulbar pain, and may be seen clinically to have evidence of occlusive vasculopathy, including retinal signs of venous stasis, hemorrhage and edema. These lesions are found histopathologically to more closely resemble anaplastic astrocytoma or glioblastoma multiforme, with extreme cellular pleomorphism, nuclear hyperchromaticity and scattered mitoses [40].

Management may involve observation in asymptomatic lesions or those without evidence of extension into the intracanalicular or chiasmal optic nerve [40]. Other treatment modalities include chemotherapy, often via combination vincristine and carboplatin therapy [41]. Targeted molecular therapy with mitogen-activated protein kinase pathway inhibitors, and anti-VEGF therapy can show variable response rates. Conventional radiotherapy has been used effectively, but with significant side effect profile including secondary brain tumors, occlusive cerebral vasculopathy, pituitary gland dysfunction and retinopathy. Fractionated stereotactic radiotherapy may be used with lower side effect profile [42]. Surgical intervention is typically reserved for intracranial or intracanalicular involvement, or in cases of blindness with painful and/or disfiguring proptosis. In such cases a multispecialty approach involving neurosurgery and oculoplastics is often utilized to excise the intraorbital and intracanacliucular portions of the involved optic nerve.

Optic Nerve Sheath Meningioma

Optic nerve sheath meningiomas (ONSM) are benign tumors arising from the cap cells of the arachnoid sheath. Primary ONSM are most commonly found in women between ages 30-50. ONSM are believed to lead to vision loss via compression of the vascular supply and axonal transport of the optic nerve. Patients typically present with painless progressive vision loss and proptosis. Other presenting symptoms may include transient visual obscurations and diplopia. Ophthalmoscopic examination classically reveals optic disc pallor and optociliary shunt vessels. CT or MRI imaging studies show thickening of the optic nerve sheath, frequently with calcifications and 'tram-track" sign (20-50% of cases), with fine tumor extensions into the orbital fat [43].

Histopathologically these lesions are found to have a meningotheliomatous or mixed-type patterns, with psammoma bodies and hyalinized calcium deposits commonly seen [37].

Management of ONSM is initially with close observation with serial bi-annual imaging for signs of tumor extension beyond the bone orbit. Generally, these lesions are compatible with good vision for many years, are non-life threatening and unlikely to spread intracranially to the point where they produce neurologic dysfunction [44]. However, if evidence of progression, treatment with radiotherapy may be considered, though with risks of toxicity to the optic nerve, retina and pituitary gland. Stereotactic or conformal fractionated radiotherapy can minimize these risks [40]. Surgical therapy is reserved for those cases with evidence of extension and threat to the optic chiasm and contralateral optic nerve.

Medulloepithelioma

Medulloepitheliomas are rare tumors arising from cells of primitive neural tube and medullary plate, and may arise anywhere along the optic pathway. Depending on the location, these lesions may induce acute clinical findings of optic nerve head edema and proptosis in orbital cases, and with retrobulbar optic neuropathy in post-chiasmal lesions.

Imaging may show fusiform enlargement of the optic nerve, resembling optic nerve glioma [40]. Diagnosis is confirmed histopathologically with hyperchromatic nuclei and neoplastic cells arranged in tubes or cords. Tumor cells stain positively for hyaluronidase and Alcian blue. Management may involve close observation or optic nerve resection. Recurrences and metastatic spread are possible [40].

Oligodendroglioma

Oligodendrogliomas originate from the oligodendrocytes and comprise up to 12% of all intracranial tumors. These lesions most commonly arise in the central hemispheres, and rarely affect the optic nerve. Patients may present with fusiform optic nerve enlargement. Histopathologically these lesions are found to be comprised of swollen oligodendrocytes cells separated by thin stroma. There have been cases of association with orbital non-Hodgkin lymphoma [45].

Schwannoma

Optic nerve schwannomas are rare, as optic nerve myelin is produced by oligodendrocytes, rather than Schwann cells. It is thus hypothesized that these lesions to arise on the optic nerve originate from the sympathetic nerves along the optic nerve sheath [46]. The radiologic appearance of an optic nerve schwannoma resembles that of an optic nerve glioma, and biopsy is required to confirm this rare diagnosis. Management options include surgical excision and stereotactic radiosurgery.

Intracranial Optic Pathway Lesions

Meningiomas

Intracranial meningioma, particularly of the sphenoid ridge, tuberculum sellae or olfactory groove may secondarily invade the optic canal and posterior orbit and compress the optic nerve. These lesions may encircle the optic nerve and compress the central retinal vein and/or artery.

Surgical intervention is generally performed to confirm the diagnosis and/or to prevent the tumor from spreading to other intracranial structures, including the contralateral optic nerve, optic chiasm, internal carotid artery, pituitary gland and cavernous sinus [40]. Adjuvant radiotherapy can be considered.

Optic Chiasm Pathology

Chiasmal lesions can be classified as intrinsic or extrinsic. Patients with optic chiasm pathology classically present with a bitemporal hemianopia, given compression of the crossing nasal nerve fibers, responsible for temporal visual field input. Attenuated optic chiasm can be seen in ocular albinism and achiasmatism [47]. Intrinsic optic chiasmal lesions include gliomas, commonly associated with neurofibromatosis type I. Demyelination may also occur in the setting of multiple sclerosis.

Extrinsic compressive lesions include pituitary tumors, meningiomas, craniopharyngiomas and aneurysms. Pituitary adenomas are the most common cause of chiasmal syndrome in adults [47]. Craniopharyngiomas can cause superior compression on the optic chiasm. These patients can be seen to have "see-saw" nystagmus, a pendulatory eye movement with alternating elevation and intorsion of one eye with simultaneous depression and excyclotorsion of the contralateral eye. Suprasellar meningiomas may lead to progressive vision loss from chiasmal compression.

Pituitary apoplexy, resulting from acute hemorrhage and decreased blood supply, often in the setting of a preexisting sellar tumor, is a life-threatening condition that may first present with bitemporal hemianopia, coinciding with severe headaches, emesis, and occasionally neck stiffness. Patients require emergent management with corticosteroids.

Optic Tract Pathology

Given that just slightly more than half (53%) of nerve fibers from the contralateral temporal visual field cross at the chiasm, lesions posterior to the chiasm along the optic tract are seen clinically to have a contralateral afferent pupillary defect. Congenital optic tract hypoplasia is exceedingly rare [37]. Optic tracts are more commonly afflicted by acquired hemorrhage, aneurism, and demyelination.

Lateral Geniculate Body Pathology

The lateral geniculate bodies may be affected by infarction of the anterior and lateral posterior choroidal arteries. Inflammatory and neoplastic processes tracking along the optic pathways or within the thalamus can also cause dysfunction.

Optic Radiation Pathology

The optic radiations may be adversely affected by subcortical infarctions, demyelinating lesions, gliomas and metastases. Progressive multifocal leukoencephalopathy or damage from radiation may also occur [48]. Pathology affecting the temporal lobe, or "Meyer's loop", characteristically result in a superior quadrantanopia visual field loss, while parietal lobe lesions demonstrate an inferior quadrantanopia.

Occipital Lobe/Visual Cortex Pathology

Occipital atrophy from infarction or degenerative disease (e.g., Alzheimer's) may uniquely present with visual phenomena and/or hallucinations [49]. Visual field defects are often of a congruous homonymous hemianopia, with or without sparing of the macula (central vision).

Idiopathic Intracranial Hypertension

Idiopathic intracranial hypertension (IIH) is characterized by an increase in pressure within the cerebral spinal space and fluid (CSF). IIH is most commonly seen in the setting of obesity, or use of steroid, Vitamin A-derived acne medication, tetracyclines, oral contraceptive pills, lithium, and growth hormone therapy. Presenting symptoms of IIH may include headaches, worse in the supine position and sometimes associated with nausea and vomiting and pulsatile tinnitus. Increased CSF pressure around the optic nerve leads to circumferential nerve compression and compromise. Clinically patients are found to have bilateral mild visual decline, dyschromotopsia, and optic nerve head swelling. Diagnosis may be confirmed with elevated opening CSF pressure on lumbar puncture, without alternative pathology discovered on CSF cytology and culture analysis, and neuroimaging. Management is initially with weight loss, oral acetazolamide and cessation of any inciting medications. Surgical management with optic nerve sheath fenestration or neurosurgical shunt implantation may be required in cases refractory to medical treatment.

PART 4: SURGICAL MANAGEMENT OF THE OPTIC NERVE

Treatment Considerations

Treatment for optic nerve disorders is managed medically for many common disorders. Surgical treatment is considered for conditions not amenable to conservative therapy. These include both direct surgical efforts and as well as recent developments in intravascular approaches. When surgical intervention is indicated, the proximal optic nerve can be accessed via an orbital approach. More extensive intraorbital and canalicular optic nerve access can require a surgical approach by a multi-specialty team.

Orbital Surgical Approach

Intraorbital optic nerve pathology may be surgically addressed from an orbital approach. Compressive optic neuropathy, such as in the setting of thyroid disease, may necessitate orbital bony expansion for the purpose of decompression. A lateral orbitotomy via a swinging eyelid, with release of the lateral canthus and inferior canthal tendon, or lateral upper eyelid crease approach, may be performed to gain access to the lateral orbital wall. Lateral orbital wall decompression may ensue via bony debulking of the greater wing of the sphenoid, with or without lateral orbital rim removal and/or advancement. The medial orbit can be accessed via a transcaruncular or endoscopic endonasal approach, the later often performed in conjunction with otolaryngology, for bony decompression or optic nerve exposure. Additional orbital decompression may be performed with orbital floor removal, though with anterior orbital strut left intact so to reduce chance of post-operative globe misalignment. Intraconal fat is often debulked for further decompression [50] typically from an infratemporal orbital approach.

The intraorbital optic nerve can additionally be accessed for biopsy or nerve sheath decompression via an orbital approach. Traditionally, optic nerve access was obtained via a medial orbitotomy with disinsertion of the medial rectus muscle [51], whereas a lateral orbital approach provides less direct access to the optic nerve due to the superior orbital fissure. More recently a superomedial upper eyelid crease approach has been utilized, avoiding need for muscle disinsertion [52]. The medial half of the upper eyelid crease is incised, and orbital septum entered. With the levator aponeurosis medial horn retracted laterally, scissors may be used to bluntly dissect within the orbital fat until the optic nerve is identified. From this approach, an optic nerve sheath fenestration or biopsy may be safely performed. Isolated optic canal pathology may also be accessed via an endoscopic trans ethmoidal approach. These techniques yield visualization limited to the intraorbital segment of the optic nerve, and thus more extensive lesions must be addressed from an intracranial approach.

Most (73%) of surgeons through the American Society of Ophthalmic Plastic and Reconstructive Surgery perform a window (i.e., square) fenestration rather than a slit [53].

Intracranial Surgical Approach

Optimal surgical access to the orbital apex and non-orbital optic nerve requires a transcranial approach by a multi-specialty team.

Frontal Transcranial Orbitotomy

In a frontal transcranial orbitotomy, a frontal bone flap, including the superior orbital rim and anterior portion of the orbital roof are removed, exposing the periorbita of the superior orbit. The frontal lobe is elevated, and the posterior portion of the orbital roof can be further debulked using a rongeur or drill. The periorbital and dura are exposed and entered. This technique may be used for debulking of an extensive sphenoid wing meningioma, orbitocranial tumor, or for removal of the intraorbital and intracanalicular optic nerve segments in setting of an enlarging optic nerve meningioma or glioma [54]. The bone flap is returned to position. It is not uncommon to see transient post-operative ptosis and extraocular motility disturbance following surgery. Other side effects include enophthalmos, pulsatile proptosis and temporalis wasting, which may mandate delayed reconstruction.

A modified eyebrow craniotomy may also function as a mini frontal craniotomy. In this technique, a 3-4 cm supraciliary brow incision is made to expose the frontal bone, and a small bone window excised. This approach provides minimal visualization of the orbital apex and thus has limited efficacy in optic nerve pathology management.

Pterional Craniotomy

A pterional craniotomy is most commonly used to obtain access to the optic canal. It is named after the pterion, the junction point of 4 bones within the skull (frontal, temporal, greater wing of sphenoid, parietal). This technique is used most commonly in removal of sphenoid wing meningioma, and allows access to the superior orbital fissure, optic strut, anterior clinoid process, roof and lateral canal walls for optic canal decompression.

In this technique the patient's head is elevated in a reverse Trendelenburg position, rotated 45 degrees to the opposite side and extended 20 degrees, secured with a three-pin skull fixation device. A curvilinear skin incision is made behind the hairline from the mid scalp to the top of the ear. The temporalis muscle is dissected from the bone and skin-muscle flap retracted. A frontotemporal bone flap is removed. The lesser wing of the sphenoid is accessed and removed to the latera edge of the superior orbital fissure. The dura may then be opened and frontal lobe elevated to allow for full access to the optic canal and nerve [55].

Optic Nerve Research

Primary research efforts on optic nerve science have centered on neuroregeneration. The disability and quality of life issues that attend blindness in humans is prohibitive to functional existence in most social environments. The number of people functionally blind in the world today is estimated at 39 million and in the United States alone is 1.028 million according to the Centers for Disease Control [56, 57].

Regeneration of the optic nerve has been studied in invertebrates for decades. The mammalian central nervous system has not seen benefits due to the disparate anatomies and physiologies of these research models. Repair of the optic nerve has focused on the intra orbital to chiasmatic segment. The various pathologies and injuries listed previously all have different pathophysiologies to consider. The principle cause of blindness is glaucoma and macular degeneration. Research in these afflictions have focused in the degenerative and inflammatory aspects of their respective disease processes [58, 59].

The cascade of neurologic events following optic nerve dysfunction represents a host of obstacles to overcome to restore function. After disruption of the optic nerve, apoptosis of the retinal ganglion cells (RGC) occurs, resulting in loss of the first order neurons and essential photoreception [59, 60]. Downstream neurological dysfunction also occurs through subsequent geniculae and ultimately the cerebral cortex-essentially an end organ loss, although rat studies show more plasticity near this end of the visual pathway [61]. Each of these structures is required to coordinate vision into a meaningful perception. Neuroregenerative efforts and stem cell research has been proposed as a way to re-animate these pathways.

Current research suggest that MSCs (mesenchymal stem cells) provide trophic factors such as chemokines, NGF, VEGF and platelet derived growth factors [62, 63, 64]. A main barrier to regeneration of the optic nerve is thought to be gliosis and inflammation consequent to injury [65, 66]. The effects of MSCs in laboratory models suggest that the regenerative effects are sustained and substantive. Functional recovery after optic nerve repair has been less robust, however. Studies in humans remain uncommon [67].

REFERENCES

[1] Neuroscience online. Contents © 1997-Present - McGovern Medical School at UT Health Department of Neurobiology and Anatomy - Site webmaster: nba.webmaster@uth.tmc.edu. Chapter 15: *Visual Processing: Cortical Pathways*Valentin Dragoi, PhD, Department of Neurobiology and Anatomy, McGovern Medical School

[2] Fundamentals and Principles of Ophthalmology. *Basic and Clinical Science Course American Academy of Ophthalmology.*, 2016.

[3] Newell, F. *Ophthalmology Principles and Concepts* 4th Edition, St Louis, 1978, The C.V. Mosby Company.

[4] Maheshwary, AS; Ross-Cisneros, FN; Carelli, V; Salomao, SR; Berezovsky, A; Moraes-Filho, MN; Moraes, MN; Belfort, R; Jr. Sadun, AA. Demonstration of Efferent Fibers in the Human Optic Nerve. ARVO Annual Meeting Abstract., May 2007.

[5] Hayreh, SS. The Blood Supply of the Optic Nerve Head and the Evaluation of it — Myth and Reality. *Prog Retin Eye Res*, 2001, 20(5), 563-93.

[6] Stanley, JA; Raise, GR. The Swinging Flashlight Test to Detect Minimal Optic Neuropathy. *Arch Ophthal*, 1968, 80, 769-771.

[7] Gala, F. Magnetic resonance imaging of optic nerve. *Indian J Radiol Imaging.*, 2015, 25(4), 421-438.

[8] Glaucoma. *Basic and Clinical Science Course.* American Academy of Ophthalmology. 2016.

[9] Newman, N; Biousse, V. Hereditary optic neuropathies. *Eye.*, 2004, 18, 1144–1160.

[10] Straatsma, BR; Foos, BY; Heckenlively, JR; Taylor, GN. Myelinated retinal nerve fibers. *American Journal of Ophthalmology.*, 1981, 91, 25-38.

[11] Lenhart, PD; Desai, NK; Bruce, BB; Hutchinson, AK; Lambert, SR. The role of magnetic resonance imaging in diagnosing optic nerve hypoplasia. *American Journal of Ophthalmology.*, 2014, 158, 1164-1171.

[12] Brodsky, MC. The "pseudo-CSF" signal of orbital optic glioma on magnetic resonance imaging: A signature of neurofibromatosis. *Survey of Ophthalmology.*, 1993, 38, 213-218.

[13] Berk, AT; Saatci, AO; Erçal, MD; Tunç, M; Ergin, M. Ocular findings in 55 patients with Down's syndrome. *Ophthalmic Genetics.*, 1996, 17(1), 15-19.

[14] Weinreb, RN; Aung, T; Medeiros, FA. The Pathophysiology and Treatment of Glaucoma: A Review. *JAMA.*, 2014, 311(18), 1901–1911.

[15] Biousse, V; Newman, NJ. Ischemic Optic Neuropathies. *New England Journal of Medicine*, (2015), 372(25), 2428–2436.

[16] Behbehani, R. Clinical approach to optic neuropathies. *Clin Ophthalmol.*, 2007, 1(3), 233-246.

[17] Kline, LB; Kim, JY; et al. Radiation optic neuropathy. *Ophthalmology*, 1985, 92(8), 1118–26.

[18] Lessell, S. Friendly fi re: neurogenic visual loss from radiation therapy. *J Neuroophthalmol*, 2004, 24, 243–50.

[19] Hickman, SJ; Miszkiel, KA; Plant, GT; et al. The optic nerve sheath on MRI in acute optic neuritis. *Neuroradiology*, 2005, 47, 51–5.

[20] Park, K; Tanaka, K; Tanaka, M. Uhthoff's phenomenon in multiple sclerosis and neuromyelitis optica. *Eur Neurol.*, 2014, 72, 153–156.

[21] Optic Neuritis Study Group. Multiple sclerosis risk after optic neuritis: final optic neuritis treatment trial follow-up. *Arch Neurol.*, 2008, 65, 727–732.

[22] Beck, RW; Cleary, PA. Optic neuritis treatment trial. One-year follow-up results. *Arch Ophthalmol.*, 1993 Jun, 111(6), 773-5.

[23] Lee, AG; Lin, DJ; et al. Atypical features prompting neuroimaging in acute optic neuropathy in adults. *Can J Ophthalmol*, 2000, 35, 325–30.

[24] Paul, F; Jarius, S; Aktas, O; Bluthner, M; Bauer, O; Appelhans, H; Franciotta, D; Bergamaschi, R; Littleton, E; Palace, J; Seelig, HP; Hohlfeld, R; Vincent, A; Zipp, F. Antibody to aquaporin 4 in the diagnosis of neuromyelitis optica. *PLoS Med.*, 2007 Apr, 4(4), e133.

[25] Borisow, N; Mori, M; Kuwabara, S; Scheel, M; Paul, F. Diagnosis and Treatment of NMO Spectrum Disorder and MOG-Encephalomyelitis. *Front Neurol.*, 2018, 9, 888. Published 2018 Oct 23.

[26] Kidd, DP; Burton, BJ; Graham, EM; Plant, GT. Optic neuropathy associated with systemic sarcoidosis. *Neurol Neuroimmunol Neuroinflamm.*, 2016, 3(5), e270. Published 2016 Aug 2.

[27] Saadoun, D; Bodaghi, B; Bienvenu, B; Wechsler, B; Sene, D; Trad, S; Abad, S; Cacoub, P; Kodjikian, L; Sève, P. Biotherapies in inflammatory ocular disorders: interferons, immunoglobulins, monoclonal antibodies. *Autoimmun Rev.*, 2013, 12(7), 774–783.

[28] Rose, GE; Vahdani, K. Optic Nerve Stretch Is Unlikely to Be a Significant Causative Factor in Dysthyroid Optic Neuropathy. *Ophthalmic Plast Reconstr Surg.*, 2020 Mar/Apr, 36(2), 157-163.

[29] Anderson, RL; Tweeten, JP; Patrinely, JR; et al. Dysthyroid optic neuropathy without extraocular muscle involvement. *Ophthalmic Surg.*, 1989, 20, 568–74.

[30] Edwards, P; et al., Final results of MRC CRASH, a randomised placebo-controlled trial of intravenous corticosteroid in adults with head injury-outcomes at 6 months. *Lancet*, 2005, 365(9475), p. 1957-9.

[31] Roelofs, KA; Starks, V; Yoon, MK. Orbital Emphysema: A Case Report and Comprehensive Review of the Literature. *Ophthalmic Plast Reconstr Surg.*, 2019 Jan/Feb, 35(1), 1-6.

[32] Farazdaghi, MK; Katowitz, WR; Avery, RA. Current treatment of optic nerve gliomas. *Curr Opin Ophthalmol.*, 2019 Sep, 30(5), 356-363.

[33] Listernick, R; Ferner, RE; Liu, GT; et al. Optic pathway gliomas in neurofibromatosis-1: controversies and recommendations. *Ann Neurol*, 2007, 61, 189–98.

[34] Kornreich, L; Blaser, S; Schwarz, M; et al. Optic pathway glioma: correlation of imaging findings with the presence of neurofibromatosis. *AJNR Am J Neuroradiol*, 2001, 22, 1963–9.

[35] Miller, N. Primary tumours of the optic nerve and its sheath. *Eye*, 18, 1026–1037, (2004).

[36] Brodsky, MC. *Pediatric Neuroophthalmology*. 3rd ed. New York: Springer, 2016.

[37] Alkatan, H; Alshowaeir, D; Alzahem, T. Optic Nerve: Developmental Anomalies and Common Tumors. In: *Optic Nerve*. Felicia Ferreri. *IntechOpen.*, 2018 Nov.

[38] Stern, J; Jakobiec, FA; Housepian, EM. The architecture of optic nerve gliomas with and without neurofibromatosis. *Archives of Ophthalmology.*, 1980, 98, 505-511.

[39] Yanoff, M; Davis, RL; Zimmerman, LE. Juvenile pilocytic astrocytoma ("glioma") of the optic nerve. Clinicopathologic study of sixty-three cases. In: Jakobiec FA, editor. *Ocular and Adnexal Tumors*. Birmingham, AL: Aesculapius Publishing Co; 1978. pp. 685–707.

[40] Miller, NR. Optic gliomas: Past, present, and future. *Journal of Neuro-Ophthalmology.*, 2016, 36, 460-473.

[41] Marsh-Tootle, WL; Harb, E; Hou, W; Zhang, Q; Anderson, HA; Weise, K; Norton, TT; Gwiazda, J; Hyman, L. For the correction of myopia evaluation trial (COMET) study group. Optic nerve tilt, crescent, ovality, and torsion in a multi-ethnic cohort of young adults with and without myopia. *Investigative Ophthalmology & Visual Science.*, 2017, 58(7), 3158-3171.

[42] Combs, SE; Schulz-Ertner, D; Moschos, D; Thilmann, C; Huber, PE; Debus, J. Fractionated stereotactic radiotherapy of optic pathway gliomas: tolerance and long-term outcome. *Int J Radiat Oncol Biol Phys.*, 2005, 62(3), 814-819. doi:10.1016/j.ijrobp.2004.12.081.

[43] Jäger, HR; Miszkiel, KA. Pathology of the Optic Nerve. *Neuroimaging Clinics of North America*, (2008), 18(2), 243–259. doi:10.1016/j.nic.2007.10.001.

[44] Egan, RA; Lessell, S. A contribution to the natural history of optic nerve sheath meningiomas. *Arch Ophthalmol*, 2002, 120, 1505–1508.

[45] Hedges, TR. III. Tumors of neuroectodermal origin. In: Miller NR, Newman NJ, Biousse V, Kerrison JB, editors. Walsh and Hoyt's *Clinical Neuro-Ophthalmology*. 6th ed. Philadelphia, PA: Lippincott Williams & Wilkins, 2005, pp. 1439-1442.

[46] Anderson, DR; Hoyt, WF. Ultrastructure of intraorbital portion of human and monkey optic nerve. *Arch Ophthalmol*, 1969, 82, 506–530.

[47] Foroozan, R. Chiasmal syndromes. *Current Opinion in Ophthalmology*, (2003), 14(6), 325–331.

[48] Jager, HR. Loss of vision: imaging the visual pathways. *Eur Radiol*, 2005, 15, 501–10.

[49] Holroyd, S; Shepherd, ML; Downs, JH. Occipital Atrophy is Associated with Visual Hallucinations in Alzheimer's Disease. *J Neuropsychiatry Clin Neurosci.*, 2000, 12(1), 25-28.

[50] Olivari, N. Transpalpebral decompression of endocrine ophthalmopathy (Graves' disease) by removal of intraorbital fat: experience with 147 operations over 5 years, *Plast Reconstr Surg*, 87, 627, 1991.

[51] Sergott, RC; Savino, PJ; Bosley, TM. Modified Optic Nerve Sheath Decompression Provides Long-term Visual Improvement for Pseudotumor Cerebri. *Archives of Ophthalmology*, (1988), 106(10), 1384–1390.

[52] Pelton, RW; Patel, BCK. Superomedial Lid Crease Approach to the Medial Intraconal Space. *Ophthalmic Plast and Recon Surg.*, 2001, 17(4), 241-253.

[53] Sobel, RK; Syed, NA; Carter, KD; Allen, RC. Optic Nerve Sheath Fenestration: Current Preferences in Surgical Approach and Biopsy. *Ophthalmic Plast Reconstr Surg.*, 2015, 31(4), 310-312.

[54] Nerad, JA. Techniques in Ophthalmic Plastic Surgery with DVD. *A Personal Tutorial.*, 2009. Elsevier: London.

[55] Eggert, HR. Pterional Approach for Microsurgical Decompression of the Optic Nerve. In: *Compressive Optic Nerve Lesions at the Optic Canal.* Springer, Berlin, Heidelberg., (1989).

[56] Pascolini, D; Mariotti, SPM. Global estimates of visual impairment: 2010. *British Journal Ophthalmology* Online First published, December 1, 2011 as 10.1136/bjophthalmol-2011-300539.

[57] Wittenborn, John S; Rein, David B. *The Future of Vision: Forecasting the Prevalence and Cost of Vision Problems.* NORC at the University of Chicago. Prepared for Prevent Blindness, Chicago, IL. June 11, 2014.).

[58] Dahlmann-Noor, AH; Vijay, S; Astrid Limb, G; Khaw, PT. Strategies for optic nerve rescue and regeneration in glaucoma and other optic neuropathies. *Drug Discovery Today.*, 2010 April, 15(7-8), 287-299.

[59] Chang, EE; Goldberg, JL. Glaucoma 2.0: neuroprotection, neuroregeneration, neuroenhancement. *Ophthalmology.*, 2012 May, 119(5), 979-86.

[60] Isenmann, S; Wahl, C; Krajewski, S; Reed, JC; Bähr, M. Up-regulation of Bax Protein in Degenerating Retinal Ganglion Cells Precedes Apoptotic Cell Death after Optic Nerve Lesion in the Rat. *European Journal of Neuroscience.*, 2006 April, 9(8), 1763-1772.

[61] Macharadze, T; Pielot, R; Wanger, T; et al. Altered neuronal activity patterns in the visual cortex of the adult rat after partial optic nerve crush—a single-cell resolution metabolic mapping study. *Cerebral Cortex.*, 2012, 22(8), 1824–1833.

[62] Johnson, TV; Bull, ND; Martin, KR. Identification of barriers to retinal engraftment of transplanted stem cells. *Invest Ophthalmol Vis Sci.*, 2010, 51, 960–70.

[63] Mead, B; Logan, A; Berry, M; et al. Paracrine-mediated neuroprotection and neuritogenesis of axotomised retinal ganglion cells by human dental pulp stem cells: comparison with human bone marrow and adipose-derived mesenchymal stem cells. *PLoS One.*, 2014, 9, e109305.

[64] Johnson, TV; DeKorver, NW; Levasseur, VA; et al. Identification of retinal ganglion cell neuroprotection conferred by platelet-derived growth factor through analysis of the mesenchymal stem cell secretome. *Brain.*, 2014, 137, 503–19.

[65] Laroni, A; Novi, G; Kerlero de Rosbo, N; et al. Towards clinical application of mesenchymal stem cells for treatment of neurological diseases of the central nervous system. *J NeuroImmune Pharmacol.*, 2013, 8, 1062–76.

[66] Giunti, D; Parodi, B; Usai, C; et al. Mesenchymal stem cells shape microglia effector functions through the release of CX3CL1. *Stem Cells.*, 2012, 30, 2044–53.

[67] Mesentier-Louro, LA; Teixeira-Pinheiro, LC; Gubert, F; et al. Long-term neuronal survival, regeneration, and transient target reconnection after optic nerve crush and mesenchymal stem cell transplantation. *Stem Cell Res Ther.*, 2019 April, 10, 121.

In: Cranial Nerves
Editor: Thomas M. Yi

ISBN: 978-1-53618-823-3
© 2021 Nova Science Publishers, Inc.

Chapter 3

CRANIAL NERVE VII: ANATOMY, FUNCTION AND CLINICAL SIGNIFICANCE

Akhil Surapaneni, BS and Craig Kemper, MD
Department of Neurosurgery, Dell Medical School, Austin, TX, US

ABSTRACT

The seventh cranial nerve (CNVII), the facial nerve, contains somatic motor, visceral motor, special sensory, and general sensory fibers. CNVII controls the muscles of facial expression along with the posterior belly of the digastric, stylohyoid, and stapedius. It gives parasympathetic innervation to several salivary glands and the nasal mucosa and provides taste sensation from the anterior two-third of the tongue. The superior portion of the facial nucleus receives bihemispheric innervation whereas the inferior portion only receives contralateral input. Second order neurons from the facial nucleus join and represent the main bulk of the facial nerve. The nervus intermedius, a branch of the facial nerve, carries general and special sensory fibers. The facial nerve exits at the superior olivary sulcus and joins a nerve bundle made up of the superior and inferior vestibular nerves along with the auditory nerve before entering the internal auditory meatus of the facial canal in the petrosal portion of the temporal bone. The course through the temporal bone is a complex route and predisposes CNVII to various pathologies unique to this anatomy.

Facial nerve palsy can occur due to lesions along its entire anatomy. Traumatic etiologies include injuries during delivery, surgical trauma, penetrating parotid injuries, facial and temporal bone fractures. Numerous inflammatory and infectious etiologies contribute to facial nerve palsies. While primary tumors of the facial nerve are rare, neoplastic compression by vestibular schwannomas and other cerebellopontine angle tumors is more common. Perineural spread of malignant tumors can occur anywhere along the course of CNVII. Metabolic, inflammatory and infectious disorders are also considered in the pathology of CNVII.

Clinical evaluation of a patient with a facial palsy begins with a history and physical examination. Timeline, disease progression, associated symptoms, and past medical history guide the examiner towards a diagnosis. The examiner should pay close attention to the external auditory canal and the surrounding structures. The House-Brackmann scale can be used to evaluate the degree of facial nerve paralysis. For patients with total paralysis of the facial nerve, electromyography and electroneurography are used to quantitate innervation to facial muscle and its response to stimulation. High resolution MRI and CT scanning can diagnose and evaluate facial nerve disorders as well as guide perioperative planning and assess the resolution of the lesion.

Following traumatic injury to the facial nerve, surgery or medical therapy alone can improve facial nerve function. In cases of nerve impingement by surrounding structures, decompression can be performed whereas interpositional nerve grafting is used for transecting nerve injuries. Oculoplastic approaches improve and maintain ocular function in facial nerve dysfunction. Facial plastic approaches are frequently used in reanimation procedures to improve function and cosmesis.

ANATOMY OF THE FACIAL NERVE

The facial nerve is composed of approximately 10,000 fibers. [1] Four brainstem nuclei contribute axons to CNVII: the motor nucleus of the facial nerve, the superior salivary nucleus, nucleus solitarius, and the trigeminal sensory nucleus. The precentral gyrus gives off somatic motor fibers that synapse in the facial motor nucleus in the caudal pons. The superior portion of the facial motor nucleus receives innervation from both hemispheres of the cortex whereas the inferior portion only receives contralateral innervation. Thus, cortical, unilateral lesions causing facial nerve palsy will spare the forehead muscles, as those receive bilateral innervation. This feature distinguishes central lesions from peripheral lesions, which will also cause weakness to the muscles in the forehead. The facial motor nucleus sends efferent fibers to the muscles of facial expression. The intramedullary fibers course around the abducens nucleus before exiting. Concurrence of lateral rectus palsy with facial palsy suggests a lesion at the level of the pons. The superior salivary nucleus sends parasympathetic fibers to sublingual/submandibular gland, lacrimal gland, and nasal mucosa. Nucleus solitarius receives special sensory fibers with gustatory information from the anterior two-thirds of the tongue. The trigeminal sensory nucleus transmits nerve fibers conveying general sensory information from a portion of the auricle, external auditory meatus, and tympanic membrane. The parasympathetic fibers and sensory fibers of the facial nerve course in the nervus intermedius, lateral to the motor fibers from the facial nucleus. The facial nerve traverses the cerebello-pontine cistern between the olive of the pons and the inferior cerebellar peduncle; this portion of the nerve is the cisternal segment.

All of these nerve fibers enter the internal auditory meatus of the facial canal in the petrosal portion of the temporal bone, along with CN VIII. The facial nerve lies in the anterosuperior portion of the facial canal. The vertical median can be identified by drawing an imaginary line from the falciform crest. The horizontal median is formed by a wedge, known as Bill's bar, the superior hemisection of the meatus. Concurrence of auditory or vestibular symptoms with facial palsy suggests a lesion in the cistern or the canaliculus. The facial nerve travels in a canal spanning 30mm in the temporal bone, making this the longest interosseous course of a cranial nerve.

The facial canal is composed of four segments: the meatal, labyrinthine, tympanic and mastoid segments. The meatal segment is the narrowest portion of the facial canal. [1] Inflammation and edema of the facial nerve may cause compression in the meatal segment, contributing to the pathogenesis of Bell's palsy. The meatal segment is also referred to as the canalicular segment. The labyrinthine segment courses between the cochlea and vestibule before synapsing in the geniculate ganglion. The labyrinthine segment has poor arterial supply, making the nerve particularly susceptible to ischemic injury at this location. The geniculate ganglion is located inside the facial canal. The tympanic branch spans about 1cm. Dehiscence of the tympanic branch of the facial canal presents in about 30% of patients, placing the facial nerve at high risk during otological procedures. [2] Distal to the geniculate ganglion, the greater petrosal nerve carries lacrimal and palatine secretory fibers to the pterygopalatine ganglion. Postganglionic fibers contain secretomotor nerves to the lacrimal gland, mucous glands of the nose, nasopharynx, paranasal sinuses, and the palate. Other branches of the facial nerve inside of the facial canal include the external petrosal nerve, nerve to the stapedius, chorda tympani, posterior auricular nerve, and nerves to the digastric and styloid muscle, respectively in a proximal to distal fashion. The external petrosal nerve sends sympathetic fibers to the middle meningeal artery. The nerve to the stapedius innervates the namesake muscle, which dampens vibrations of the stapes bone. This modulates the amplitude of sound waves in the middle ear. The chorda tympani carries sensation of taste from the anterior two-thirds of the tongue. The posterior auricular nerve branches into the auricular branch, innervating the auricularis posterior muscle and the intrinsic muscles of the ear, and the occipital branch which courses posteriorly to supply the occipitalis muscle. This nerve has a relatively long length and supplies clinically insignificant muscles, making it a potential donor candidate for a nerve graft. [3]

Lesions proximal to the geniculate ganglion affect the intra-temporal branches of the facial nerve, presenting with hyperacusis and loss of lacrimation and taste. Lesions between the geniculate ganglion and the stylomastoid foramen preserves lacrimation. Lesions distal to the stylomastoid foramen preserve lacrimation, taste and do not present with hyperacusis. Both central and distal lesions of the facial nerve can cause excessive lacrimation, in a condition known as crocodile tears. This is due to regrowth of the chorda tympani branches into the lacrimal gland. At the stylomastoid foramen, two more branches of the facial nerve course towards the digastric and stylohyoid muscles, respectively. These two suprahyoid muscles elevate the hyoid bone as well as facilitate complex jaw movements. The facial nerve exits through the stylomastoid foramen and branches into the temporal, zygomatic, buccal, marginal mandibular, and cervical nerves which innervate the muscles of facial expression.

After the facial nerve exits through the stylomastoid foramen as the pes anserinus, it branches into two trunks. The upper trunk gives off the frontal, zygomatic, and buccal nerves. The frontal branch of the facial nerve courses superiorly across the zygoma and runs parallel to the superficial temporal vessels. It innervates the occipitofrontalis muscle, corrugator supercilii, orbicularis oculi, and the anterior and superior auricularis muscles. The occipitofrontalis muscle draws the scalp backwards, causing the eyebrows to raise superiorly and the forehead to wrinkle. The corrugator supercilii brings the eyebrows together and downward, creating vertical furrows in the center of the forehead. The orbicularis oculi muscle closes the eyelids. The zygomatic nerve travels over the zygomatic arch innervating the zygomatic, orbital, and infraorbital muscles. The zygomatic muscle draws the corner of the lips superiorly and posteriorly, allowing one to smile. The buccal nerve travels anteriorly with Stensson's duct to innervate the buccal, upper lip, and nostril muscles. The buccal muscle holds the face to the cheek and assists in chewing. The lower trunk gives rise to the marginal mandibular branch and the cervical branch. The marginal mandibular branch courses inferiorly to the parotid gland, in a plane that is deep to the platysma, to muscles in the lower lip and chin. The depressor labii inferioris, which depresses the lower lip, depressor anguli oris, which depresses the corners of the mouth, and mentalis muscle, which furrows the skin on the chin and elevates the lower lip, are innervated by this nerve. The cervical branch also follows the same plane as the marginal mandibular branch and it innervates the platysma itself. While the peripheral branches of the facial nerve act relatively independently, there are interconnections especially between the zygomatic and buccal branches. While this improves recovery in an isolated injury to a branch, it also facilitates synkinesis during recovery.

DEVELOPMENT

Development of the facial nerve starts in the first month of development. The acousticofacial primordium originates from the otic placode, which is the primordial structure of the ear. Topologically, it is located in the r4 rhombomere. Rhombomeres are distinct developmental entities in the rhombencephalon, which are transcriptionally independent. Each rhombomere and its corresponding pharyngeal arch are regulated by the same group of HOX genes. [4] The acousticofacial primordium gives rise to the trunks of the facial and vestibulocochlear nerves. Near the second month of development, the geniculate ganglion forms in the second pharyngeal arch, and for that reason it will end up supplying all of the structures from the same arch. The acousticofacial primordium separates into a rostral and caudal portion. The caudal portion descends to synapse with the geniculate ganglion, making the main facial nerve trunk. The rostral portion becomes associated with the first pharyngeal arch, eventually becoming the chorda tympani nerve. By the seventh week, the chorda tympani runs with the V3 branch of the trigeminal nerve in the lingual nerve, providing taste sensation to the anterior two-thirds of the tongue. The motor nucleus of the seventh nerve develops around the sixth week, separately from the nerves in the nervus intermedius and geniculate ganglion. It forms in the middle ear between the membranous labyrinth, a structure derived from the otic placode, and the stapes, a derivative of the second pharyngeal arch. It passes its fibers into the mesenchyme of the second pharyngeal arch, under regulation by HOX genes. During this time, the greater petrosal nerve and the extratemporal branches of the facial nerve also develop from the main trunk of the facial nerve.

By the end of the eighth week, a capsule made of cartilage encircles the facial nerve, stapes muscle, and stapedial artery. This is the early facial canal. By the 12th week, the muscles of the face develop and receive innervation from the facial nerve. If there is a mismatch between branches of the nerve and the muscle that it innervates, the muscle atrophies and undergoes transformation into adipose and fibrous tissue. This finding is seen in Moebius syndrome, a congenital sixth and seventh nerve palsy.

CLINICAL IMPACT

Facial paralysis can impair the patient's ability to express basic emotions. [5] For example, characteristic facial expressions are produced when eating especially bitter or sweet foods, an evolutionary development that conveys valuable information to a group, [6] specifically, high levels of synkinesis is associated with deficits in expressing happiness. Eyelid weakness and brow ptosis causes problems in showing sadness. Overall slowing of facial movements is perceived as a delayed reaction to stimuli, making it difficult for the patient to show surprise. During childhood development, the emphasis that is placed on interpreting facial expressions shows the paramount importance of facial expression. Infants learn to read eye gaze in the first few months of life and consequently, respond with a smile when an adult's eye gaze is directed towards them. [7] Furthermore, facial movement is important in nonverbal communications. For example, eyebrow movements are so influential in communication, that sign language communicators use them to express nuances in their hand signs. Facial motion is a key variable used by observers to identify faces, suggesting that facial identity is an integration of the invariant and mutable aspects of facial structure. [8] The production and perception of facial expressions is a cognitive process, and thus facial palsy can also be seen by an observer as a cognitive impairment. [9]

Given the widespread influence of facial expression in daily functioning, facial paralysis is a significant disability. Patients with facial nerve palsy report lower social functioning due to their physical impairments. [5] Up to 65% of those with facial neuromotor disorders also reported depressive symptoms. [10] The inability to smile alone is associated with an increased risk of depressive symptoms, suggesting the motor action itself reinforces positive emotions.

EVALUATION OF FACIAL NERVE LESIONS

The House-Brackmann scale is a clinical tool that is used to evaluate the degree of facial nerve paralysis. It is scored from 1 to 6, based on the level of paralysis in the eye, mouth, and forehead both at rest and during motion. The Sunnybrook Facial Grading system is another popular clinical tool, which has comparable validity with other grading systems. [11, 12] and comprises three domains: resting facial symmetry, voluntary motion, and degree of synkinesis during motion. The eye, mouth and cheek region are observed with the face at rest. Forehead wrinkle, gentle eye closure, open mouth smile, snarl, and lip pucker are the facial movements that are evaluated for symmetry and synkinesis. Another scale, the electronic clinician graded facial function scale (eFACE), allows clinicians to compare a patient's function to a visual analogue. This tool allows clinicians to better standardize the severity of a patient's presentation. Patient reported outcome measures, PROMs, can supplement the clinical evaluation of facial palsy because they capture its impact on a patient's quality of life. The Facial Clinimetric Evaluation (FaCE) is a validated and commonly used PROM which looks at the impact of paralysis for different facial muscles on the psychosocial and physical aspects of daily life. [13] The synkinesis assessment questionnaire is a validated PROM to determine the amount of synkinesis. [14] The facial disability index is a self-reported patient questionnaire which aims to quantify the degree of disability caused by facial nerve paralysis. [15]

Electrophysiological assessments can objectively quantify the physiological extent of nerve damage. Motor nerve conduction studies (NCS) measure properties of a stimulus-response system, with an electrical stimulus being applied to a nerve and the response being measured with electrodes over the belly of the muscle of interest. [16] Clinically relevant properties include the latency from the stimulus to the onset of the response and its amplitude. The early signal that is measured is the 'M' response, which represents orthodromic signal conduction from the stimulation. The latter signal, called the 'F' response occurs from antidromic signal transmission from the stimulus to the anterior motor horn and then orthodromically to the muscle. This pathway, referred to as the H-reflex, provides some information about conduction through the anterior horn cell. The nerve can be stimulated in multiple sites, and the latency of response from the proximal site can be subtracted from the latency of the distal site. When the latency difference is divided by the distance between the two sites of stimulation, the conduction velocity of the nerve is obtained. The conduction velocity through sensory nerves can be measured directly by applying the stimulus and measuring the response on the same nerve.

In electromyography (EMG), a recording needle is placed directly into the muscle, and recordings are taken at rest, light activity, and maximal activation. The signal for muscle contraction is transmitted from the anterior motor horn, through the nerve's axon, to a group of muscle fibers. The electrical potential in the muscle fibers that are depolarized by a single axon is known as the motor-unit potential. Normal skeletal muscle will display no spontaneous electrical activity at rest. All of the muscle fibers fire in synchrony, producing a monophasic signal. However, denervated muscle at rest has been shown to display positive, sharply peaked waves and fibrillations, often with two or three phases. Diseases involving the nerve or the muscle will cause the fibers to fire asynchronously, producing a polyphasic signal. In order to increase the strength of contraction, more motor-units are recruited and the frequency of depolarization of the fibers in each motor unit increases. In myopathies, the number of motor units remain the same, but with exertion, the number of motor units that are recruited increases dramatically. In peripheral nerve injuries, the number of motor units is decreased, so the frequency of depolarization increases to compensate.

EMG and NCS can be used to distinguish between supranuclear and peripheral facial nerve injuries. NCS can show sharp waves and fibrillations, consistent with denervation, in some types of peripheral nerve injuries, but not in supranuclear injuries. Furthermore, it can be used to distinguish between axonal neuropathies and demyelinating diseases. Axonal neuropathies will demonstrate reduced motor unit potentials and show signs of denervation. However, the conduction velocity will be normal. In contrast, demyelinating diseases have reduced conduction velocities and even a conduction block, in severe injuries. However, signs of denervation will be absent because the axons are still intact. Electrophysiological studies can also be used to determine the extent of peripheral nerve injury. In neuropraxia, reduced motor unit potentials are seen but no denervation potentials. In axonotmesis, motor and sensory potentials distal to the lesion will be reduced in amplitude. Positive sharp waves and fibrillation potentials are also seen. On follow-up exam, evidence of reinnervation (return of polyphasic signals) is seen. Neurotmesis initially presents like axonotmesis on electrophysiology, but re-innervation is incomplete.

Imaging

CT and MRI have complementary roles in evaluating facial nerve lesions. CT imaging is superior to MRI in evaluating damage to osseous structures, such as fracture, degeneration, periosteal reactions, and metabolic bone disease. Temporal bone CT can evaluate lesions of the internal auditory canal and its patency. [17] CT has the advantage of visualizing the relationship of the facial nerve to normal bony anatomy, such as the middle ear bones. [17] Supranuclear lesions are best evaluated using standard MRI. Nuclear lesions of the facial nerve require a T2-weighted sequence to visualize the entire brainstem. Visualization of the cisternal, meatal, and labyrinthine segments of the facial nerve requires the combination of T1 and T2-weighted MRI sequences. MRI can visualize small perineural tumors and ischemic injury to these locations. In the temporal bone, MRI can visualize neoplastic and inflammatory processes that affect the nerve. Gadolinium-enhanced MRI is useful for distinguishing neoplastic from inflammatory lesions of the facial nerve. [18] The portion of the facial nerve within the facial canal that is distal to the geniculate ganglion, the tympanic and mastoid segments, is best evaluated by both CT and MRI. The extracranial portion of the facial nerve can be visualized by MRI.

Diffusion tensor imaging is an MRI technique that assesses the diffusivity of water across axonal tracts. It has high sensitivity in detecting CPA tumors and perineural spread of extracranial tumors. [19, 20] As a general rule, all patients who have facial paralysis should be evaluated with an intracranial MRI, and those who have damage secondary to trauma should also receive a temporal bone CT.

The facial nerve can enhance in the absence of pathology on contrast MRI imaging, making it difficult to identify pathology. [21, 22] Normal enhancement occurs between the labyrinthine segment and the stylomastoid foramen due to the capillary plexus that surround the nerve. However, enhancement of the extratemporal branches of the nerve should be interpreted as abnormal. [22]

FACIAL NERVE PALSIES

After a nerve is injured, it gradually loses its function. If the edema is minimal and the nerve is anatomically intact, a conduction blockade (neuropraxia) will be the only result. As the degree of edema becomes more significant, the nutrient supply to the damaged nerve becomes more restricted, resulting in axonal death (axonotmesis). When this occurs, the axon degenerates in a retrograde fashion to the meatus of the facial canal. However, the endoneurium remains intact, allowing the axon to regrow in its sheath. If the endoneurium also becomes damaged (neurotmesis) during the course of injury, there is a risk for the development of synkinesis. Facial synkinesis is defined as abnormal movements or combination of movements of muscles during voluntary or spontaneous facial expressions. An example of synkinesis is involuntary eyelid closure when attempting to smile. The leading etiological theory suggests that the nerve fibers grow haphazardly during regeneration after injury, incorrectly rewiring the face. The ephaptic transmission hypothesis is an alternative theory, suggesting that synkinesis occurs when poorly re-myelinated motor neurons "cross-talk" by generating electrical field potentials that alter the excitability of nearby neurons.

Idiopathic Facial Palsy

Bell's palsy is an idiopathic paralysis of the peripheral facial nerve. It accounts for about half of cases of facial weakness. [23] Bell's palsy usually presents unilaterally, with a sudden onset, and generally resolves in a few months. The rapidly progressive paralysis distinguishes it from neoplastic nerve compression. The ipsilateral face begins to droop, most prominently visualized by looking at the corner of the mouth and the eyebrow. Additionally, there is a flattening of the nasolabial fold in affected patients. Bell's phenomenon, describing the upward deviation of the eye on attempted closure of the eyelid, is pathognomonic for Bell's palsy. [24]

Both the upper and lower parts of the face are affected, differentiating it from supranuclear lesions causing facial nerve palsy. The etiology of Bell's palsy has been heavily debated over the years. One theory suggests that it is a peripheral demyelinating disease, similar to Guillain-Barre disease. [25, 26] Reactivation of latent herpes virus infection along the facial nerve could play a role in initiating an inflammatory response. The subsequent inflammation causes cellular swelling and edema of the nerve, which leads to nerve compression along its course. [27–29] Additionally, there has been a link demonstrated between the intranasal flu vaccine and the development of Bell's palsy 31 to 60 days after vaccination. [30] However, the mechanism is purported to be a reactivation of herpes virus due to an immunocompromised state, instead of a direct effect of the influenza virus. While vesicular lesions in a dermatomal distribution provide evidence of herpes zoster infections, differentiating it from a diagnosis of Bell's Palsy, they may be absent in a clinical syndrome known as Zoster sin herpete. [31]

Laboratory tests and imaging are not routinely performed in Bell's palsy and are reserved for patients who have recurrent disease or fail to show any clinical improvement in three weeks. [32] It is still recommended that a neurologist or an otolaryngologist evaluate these patients to rule out other causes of weakness. [33] Lyme serologies should be obtained in endemic areas. Furthermore, while Bell's palsy rarely presents in children younger than ten years of age, Lyme serologies are positive in up to half of these cases. [34]

IV steroid and antiviral medication are the only therapies that have been shown to improve the chances of recovery. [35–42] Furthermore, they have also been shown to reduce the degree of synkinesis during recovery. The evidence shows that there is no benefit of using antivirals alone versus placebo in treating Bell's palsy. [43] However, using antivirals in addition to steroids seems to be beneficial. [44] In particular, valacyclovir seems to add additional treatment benefit with glucocorticosteroids. [39]

Patients with severe axonotmesis may benefit from surgical decompression of the facial nerve in the middle fossa. [45] The benefit of acupuncture has been explored, but the evidence suggests that it neither improves the recovery rate or morbidity associated with the disease. [46] Patients who have an incomplete paralysis usually recover completely whereas patients with House-Brackmann grade 6 paralysis may recover incompletely and have a protracted course of disease.

Traumatic Injury to the Facial Nerve

Trauma comprises the second most common etiology of facial nerve injuries, accounting for 8-22% of these cases. Blunt or penetrating trauma to the temporal bone is the most common cause of traumatic facial nerve injury. Another significant portion of facial nerve injuries occur from birth trauma, either due to pelvic abnormalities or the use of forceps in delivery. Other causes of injury are surgical trauma, penetrating parotid or middle ear traumas, barotrauma, facial fractures, and temporal bone fractures. High resolution CT can be used to evaluate the temporal bone for signs of trauma. Electrophysiologic testing can help determine the extent of nerve injury, which aids in determining the prognosis of the injury. Primary end-to-end neurorrhaphy is the preferred management for a complete transection of the nerve. Facial nerve decompression may also provide benefit in high grade nerve trauma. Secondary facial reanimation procedures may improve patients' quality of life after primary facial nerve repair fails.

Temporal bone fractures cause injury to the facial nerve in 7-10% of cases. [47–49] The pyramidal shape of the temporal bone lessens the pressure of impact, making the bone relatively difficult to injure. Thus, temporal bone fractures are typically due to high velocity mechanisms like motor vehicle collisions. Traditionally, these are characterized by the orientation of the fracture, either longitudinal or transverse to the petrous ridge of the temporal bone. [50] Transverse fractures and fractures violating the otic capsule are more likely to involve the facial nerve. [51]

Complete paralysis due to open injury warrants emergent surgical exploration. [52] The wound should be copiously irrigated and antibiotics should be administered. After identifying the proximal and distal ends of the injured nerve, primary neurorrhaphy should be performed as primary repair has the best outcomes. [53] If the primary nerve does not have sufficient length for anastomosis, then an interposition nerve graft can be performed. The great auricular nerve or sural nerve are commonly used. A finding of greater than 90% axonal degeneration on electroneurography, even several days after presentation, is an indication for surgery as well. Patients with non-regenerative axons can have an intraneural hematoma, making surgical decompression beneficial. Delayed or incomplete paralysis should be initially treated with steroids. These patients typically have a good prognosis, meaning recovery to House-Brackmann grade II or better. [54]

Infectious Facial Palsy

Ramsey Hunt syndrome is caused by the eruption of herpes zoster oticus in the external ear. It is the second most common atraumatic cause of facial nerve paralysis, with the incidence ranging from 3-20% in patients with facial nerve palsy. [55–57] Clinical features include a vesicular rash in the external ear and lower motor neuron facial palsy, as well as tinnitus, hearing loss, nausea, vomiting, vertigo, and nystagmus due to involvement of the vestibulocochlear nerve in the facial canal, and glossopharyngeal and vagal neuropathy from spread of the seventh nerve fibers in the tongue. Compared to adults, children are more likely to develop a vesicular rash after the onset of facial palsy, making the early stages of the infection difficult to distinguish from Bell's palsy. However, adults are more likely to develop associated symptoms of hearing loss, tinnitus, vertigo, and glossopharyngeal/vagal nerve symptoms. An important feature of Ramsay Hunt syndrome is a painful neuralgia, radiating to the ear, which is accompanied by lacrimation, nasal congestion, and salivation. Zoster sin herpete is diagnosed if the patient has the characteristic radicular neuropathy and PCR evidence of varicella-zoster virus (VZV), without the appearance of a vesicular rash. A small percentage of patients with Bell's palsy have laboratory evidence of a VZV infection, putting zoster sin herpete in the differential diagnosis. [58, 59]

The namesake, Dr. Ramsay Hunt, deduced that areas of somatic sensation in the ear that were spared in trigeminal neuralgia surgery matched perfectly with the distribution of vesicular lesions during an eruption of herpes zoster oticus. This led him to conclude that sensation of these areas of the outer ear were derived from the geniculate ganglion of the facial nerve, the petrous ganglion of the glossopharyngeal nerve, or from the jugular and plexiform ganglia of the vagus nerve. [60]

Patients with Ramsay Hunt syndrome have a more severe paralysis than those with Bell's palsy and are less likely to completely recover. [61] Facial function is completely lost within one week. Complete paralysis is twice as likely to occur compared to incomplete paralysis, with patients experiencing complete palsy being less likely to recover and more likely to develop synkinesis. [57] Herpes zoster is known for causing necrotic dorsal root ganglionitis. However, in Ramsay Hunt syndrome, no damage to mild inflammation with lymphocytic infiltrate is seen in geniculate ganglion. [62, 63] The facial nerve undergoes perivascular inflammation with features of demyelination and axonal loss. [64, 65] One report describes the formation of a hemorrhagic neuroma of the seventh nerve. [62] Ramsay Hunt is a clinical diagnosis, which can be confirmed by VZV

PCR, preferably from ear exudates. [66] MRI and CSF studies have a limited clinical utility for diagnosis. [67]

The standard of care in treatment for Ramsay Hunt is early initiation of prednisone and acyclovir, within the first 72 hours. [68]

Lyme disease is an infection by the spirochete *Borrelia burgdoferi* in North America, *Borrelia afzelii* in Europe, and *Borrelia garinii* in Asia, carried by the Ixodes tick. Primary infection begins with a target-like skin lesion, known as erythema migrans. If untreated early, it can spread to involve the cardiovascular, central nervous, and musculoskeletal systems. In the early disseminated stage, it can cause cranial neuropathies, especially facial nerve palsy. Facial palsy occurs in 7.5% to 10% of untreated Lyme disease cases with rapid onset bilateral flaccid paralysis. [69-71] Facial palsy is usually present as part of a polyradiculopathy, which sometimes also involves the meninges. Meningoradiculoneuritis, also called Bannwarth syndrome, manifests with severe segmental, burning pain which does not respond to pain medications. 75% of patients go on to develop focal neurological deficits. [72] Neuroborreliosis responds well to penicillin, ceftriaxone, cefotaxime, and doxycycline. [73] For patients with meningitis, cranial neuritis, and radiculitis without parenchymal CNS disease or severe symptoms, oral doxycycline is recommended. [73] Patients with parenchymal brain or spinal involvement should be treated with parenteral antibiotic therapy. Treatment with corticosteroids and antivirals is associated with worse long-term recovery of facial control. [74] Recovery of function occurs gradually over the following months, although, up to 16 to 23% of patients have residual deficits, including a mixture of hypo- and hyperactivity with synkinesis. [75]

Tuberculous otitis media is a rare cause of facial nerve palsy, especially in children. The middle ear is secondarily infected from either hematogenous spread or direct spread from a primary focus via the eustachian tube. It comprises 3-5% of cases of chronic suppurative otitis media, with the majority of cases occurring in patients younger than 15 years of age. [76, 77] The classical clinical presentation is painless otorrhea with facial paralysis and findings of bony necrosis and multiple perforations in the middle ear on otoscopy, though only few patients exhibit all of these features. [78] Specifically, facial palsy presents in around 16% of adult cases and 35% of cases involving children. [79, 80] Thus, given the wide variety of presentations, clinicians should have a low threshold for suspecting tuberculous otitis media in a patient with chronic suppurative otitis media and facial palsy. Patients have deafness out of proportion with the extent of visualized disease, on audiogram. Plain films and CT scans of the mastoid can show pockets of air in the mastoid and a chest x-ray may show signs of pulmonary tuberculosis, although neither of these findings are diagnostic nor necessary. [81] Diagnosis is made from a combination of a smear and culture of the ear discharge and histopathological examination of the middle ear, as each of the above have a low positivity. While tuberculous otitis media can be treated with the standard anti-tuberculosis antibiotic regimen, development of facial palsy is an indication for surgery to decompress the facial nerve and debridement of necrotic tissue. [82]

Peripheral facial palsy can develop during any stage of HIV infection, even preceding the formation of anti-HIV antibodies. In stages I and II it can manifest as an acute mononeuritis like Bell's palsy or as a gradual polyneuropathy, similar to Guillain-Barre. [83–85] As HIV is neurotropic, it is proposed that the virus damages neurons as the pathogenesis of facial palsy, while others argue that the inflammatory response from the infection causes neural swelling and compression of the nerve, as in Bell's palsy. [86, 87] In the later stages of HIV infection, facial palsy develops as a consequence of immunosuppression which leads to reactivation of herpes zoster. This theory is supported by clinical and laboratory evidence of zoster infection in HIV patients, although acute co-infection with zoster cannot be ruled out. [88, 89] AIDS patients with systemic lymphoma can develop skull base metastases that can cause nerve compression. [90] Lastly, peripheral palsy can be part of a widespread polyneuropathy, due to HIV-related inflammatory demyelinating polyradiculopathy. [91]

Facial palsy has also been described as a rare presenting feature in other viruses including poliovirus, mumps virus, CMV, Epstein-Barr, tuberculous leprae, bacillary angiomatosis, and dengue virus. [92–97]

Primary Neuroinflammatory Conditions

The incidence of facial palsy as a presenting feature in multiple sclerosis is 1-5%. [98–100] However, 20-50% of patients with MS develop facial palsy in their lifetime, making it a relatively common feature of the disease. [101] Facial palsy in MS occurs due to a central lesion, most commonly in the cortex and in the pons. However, the clinical presentation mimics a peripheral lesion, with weakness in all facial muscles. This raises the possibility that the palsy occurs due to the concurrence of two different disease processes. [102] In MS, facial palsy commonly occurs with other cranial neuropathies, which is a point of differentiation from Bell's palsy. Nearly all of these patients have a lesion that was identified on MRI. Patients with facial palsy who are suspected to have MS should be started on long-term therapy with disease-modifying agents, with acute exacerbations treated with corticosteroids.

Facial palsy is a common presenting sign in myasthenia gravis. Isolated facial weakness without any signs of extraocular muscle involvement is very rare, however. [103] Worsening facial weakness with exertion or a decrement in the CMAP amplitude with repetitive stimulation on EMG is characteristic of myasthenia gravis.

Facial weakness is also commonly observed in Guillain–Barré syndrome (GBS). GBS can involve either one or both facial nerves, in the setting of a cranial nerve polyneuropathy. [104, 105] Classically, limb weakness precedes facial weakness. [106] Interestingly, even though the facial nerve is involved, enhancement of the seventh nerve is not seen on MRI. [107] On EMG, slowed conduction velocities are seen. CSF analysis shows cyto-albuminic dissociation, characteristic of Guillain–Barré. Miller Fisher syndrome, a clinical subtype of Guillain–Barré with extensive cranial nerve involvement, also commonly includes facial nerve paralysis. [108] Treatment for Guillain–Barré is mostly supportive, with immunoglobulin plasmapheresis reserved for severe cases.

CONGENITAL CAUSES OF FACIAL PALSY

Hereditary hypertrophic neuropathy, one of the subtypes of Charcot-Marie-Tooth, is a rare cause of facial nerve palsy. Melkerson-Rosenthal syndrome (MRS) is a rare syndrome in which patients can have a classic triad of facial nerve palsy, orofacial edema, and lingua plicata (fissured tongue), though most patients are oligosymptomatic on presentation. [109–112] Patients have recurrent, resolving symptomatic episodes, which progressively become more severe over time. The facial nerve palsy is usually unilateral, on alternating sides of the face. There is a familial association, with multifactorial inheritance. [113] MRS is a clinical diagnosis, histopathological analysis shows multinucleated giant cells, noncaseating granulomas, and a perivascular lymphocytic infiltrate. [114] Treatment includes non-steroidal inflammatory drugs, corticosteroids, and antibiotics. [115] Facial nerve decompression has been performed for a small number of cases. [116, 117]

Moebius syndrome presents with a congenital, non-progressive sixth and seventh palsy. The incidence is estimated to be 1 in every 250,000 live births. [118] Infants with Moebius syndrome have problems with sucking, lack of facial mimicry, fixed gaze, and incomplete lid closure during sleeping. Congenital malformations in the limbs, face, and chest are also observed. The pathogenesis of Moebius syndrome is unclear, but some proposed etiologies include ischemia during pregnancy, abnormalities in rhombencephalon development, and fetal toxin exposure. [118] Treatment involves speech therapy, physical therapy, and psychomotor interventions.

FACIAL PALSY SECONDARY TO SYSTEMIC DISEASES

Facial palsy is the most common finding in neurosarcoidosis, commonly occurring in the setting of other neurological findings. [119] Neurosarcoidosis is a great mimic of other diseases, making its diagnosis very difficult. Furthermore, molecular markers like ACE level are only elevated in a quarter of patients due to lack of systemic involvement. MRI with contrast of the brain and lumbar puncture are more commonly used to diagnose neurosarcoidosis. MRI shows contrast enhancement of the meninges, cranial nerves, or HPA axis. [120, 121] Lumbar puncture reveals elevated CSF protein, pleocytosis with lymphocytic predominance, low CSF glucose, and oligoclonal bands. [120, 122–124] Definitive diagnosis is established with a biopsy, either of the CNS or systemic lesions. Corticosteroids are a good first line therapy for patients with neurosarcoidosis. Immunologics like methotrexate, mycophenolate mofetil, azathioprine, cyclosporine, and cyclophosphamide can be considered as well.

Microangiopathic diseases, like hypertension, diabetes, and hypercholesterolemia, are associated with a higher incidence of Bell's palsy due to ischemia to the vaso vasorum. [125] Other systemic and metabolic causes of seventh nerve palsy include amyloidosis, hyperthyroidism, acute intermittent porphyria, carbon monoxide toxicity, tetanus, diphtheria, vitamin A deficiency, ethylene glycol ingestion, and alcoholism.

Bilateral facial palsy is a rare clinical finding, with an incidence of one per 5 million persons. [126] and should raise suspicion for an identifiable etiology. Reported causes of bilateral facial paralysis include Lyme disease, Guillain–Barré, leukemia, sarcoidosis, infectious mononucleosis and trauma. [104, 106]

Treatment

The etiology and severity of the facial nerve paralysis plays an important role in the treatment options. The duration of the paralysis is important too, as motor end plates completely degenerate two years after onset. Temporary measures can be done if full nerve recovery is expected or as a bridge between facial reanimation and the restart of nerve conduction. Facial reanimation procedures can be divided into two categories, dynamic and static. Dynamic procedures restore a degree of control of the affected facial muscles, making them preferred over static procedures. Primary neurorrhaphy or interpositional nerve grafting can be done in the acute setting within the first several weeks. The goal of surgery should be to create a tension-free segment between the two nerves, as tension can cause nerve ischemia, which inhibits complete regeneration and increases the risk of postoperative reflex sympathetic dystrophy. [127] Coaptation of the nerve should be performed distally to the stylomastoid foramen in order to reduce synkinesis. [128, 129] Combining an interpositional nerve graft with the great auricular nerve to upper facial nerve branches with another nerve graft to the lower facial nerve branches has been shown to restore facial function without synkinesis. [130] In this acute stage, surgical decompression can be beneficial, as the nerve can still recover well.

In the setting of intermediate nerve paralysis, between three weeks to two years, primary repair is not an option because transected nerves have a tendency to retract, due to physical stretching and inflammatory degradation. If the damaged nerve does not regenerate properly by this time, primary repair will not be possible due to the great deal of axial tension on the retracted ends. Thus, nerve transfers and nerve crossover procedures are the treatment of choice, as the motor end plates are still viable. The cross-facial nerve graft (CFNG) uses branches of the contralateral, intact facial nerve to innervate the muscles of the hemiface on the injured side. [131] CFNG poses a challenge, however, because axons must travel a long distance to reach their targets on the motor end plates, all the while the end plates continue to degenerate. [132] To mitigate this challenge, a partial hypoglossal nerve transfer can be done at the same time as the CFNG, which maintains neural input to the motor end plates while the grafted facial nerve continues to grow into its territory. [53] The major risk of CFNG is causing an injury in the intact nerve, if a dominant branch of the nerve is transferred to the hemiplegic face. [133] Nerve transfers using other nerves is also possible, though the contralateral facial nerve provides the best spontaneous emotional expressivity. [134]

The hypoglossal, masseteric, spinal accessory, ansa cervicalis, recurrent laryngeal, and phrenic motor nerves are good candidates for nerve transfer, with the hypoglossal transfer being the best described. It is possible to only transfer some of the hypoglossal nerve fibers, preserving some tongue movement and swallowing function. [135–137] Either the superior or inferior half of the hypoglossal nerve is transferred at the styloid process, with similar results. [138] End-to-end hypoglossal-facial nerve anastomosis can be done for central facial palsy, at the level of the facial nucleus. [139] The nerve to the masseter can also be used for nerve transfer, possibly even with a faster recovery compared with hypoglossal nerve. [140] Patients with a masseter nerve transfer exhibit a spontaneous smile, presumably due to the physiological involvement of the masseter muscle during smiling. [141] The major risk in using the masseter nerve for nerve transfer is potential injury to intact distal branches of the facial nerve. [142] To avoid this, the nerve to the masseter can reliably be found approximately 1cm inferior to the zygoma, 3 cm anterior the tragus, and 1.5 cm deep to the superior muscular aponeurotic system. [143]

In the case of chronic facial paralysis, usually greater than two years after injury, the facial muscle has atrophied and the motor end plates have degenerated. Therefore, muscle transposition using local musculature or free muscle grafts are the mainstays of treatment. Temporalis muscle is the most common regional muscle transfer technique. The temporalis tendon is released from the coronoid process and the muscle is swiveled around and pulled over the zygomatic arch. The tendon is affixed to the oral commissure. Free tissue transfers have been described as single-stages or two-stage procedures. In a single stage procedure, the latissimus dorsi tissue flap is used, with anastomosis of the thoracodorsal nerve to the contralateral facial nerve. [144, 145] A single stage gracilis free flap procedure has also been described, with innervation from the nerve to the masseter. [146, 147] Due to the superiority of the contralateral facial nerve in restoring spontaneous facial function, the masseter nerve should be considered after cases of failed CFNG. [148, 149] The two stage procedure first begins with a CFNG and then a gracilis free tissue flap follows after patients report a tingling sensation in the injured hemiface. [150] Idiopathic nerve palsy can be treated with steroids but surgery is indicated for rapidly progressive disease. Other muscles that have been used for free-tissue transfer include the pectoralis minor muscle, free abductor hallicus, and the extensor digitorum brevis. [151–153]

Cases of bilateral facial paralysis pose a challenge due to the lack of a functional contralateral nerve. In these cases, there are several options for surgical interventions, using the nerve to the masseter for bilateral gracilis free flap transfers, bilateral end-to-side anastamoses between the hypoglossal nerve and the nerve to the gracilis, or the spinal accessory nerve for bilateral gracilis transfers.

Ocular Considerations of Facial Paresis

Eye care is important because lagophthalmos secondary to levator palpebrae superioris palsy can cause exposure keratitis. Botulinum toxin can be injected into the levator palpebrae superioris or the superior tarsal muscle, to temporarily counteract lagophthalmos. Hyalauronic acid can also be injected into the space between the levator aponeurosis and superior tarsal muscle to close the eye as well. Permanently correcting upper eyelid retraction involves gold weight insertion or less commonly with palpebral spring implantation or temporalis muscle transfer. [154]. Clinicians should screen for an ectropion by identifying an increased width of the lid aperture. Eyelid laxity in the lateral eye causing an ectropion can be corrected with a lateral tarsal strip or lateral transorbital canthopexy. Laxity in the medial side can be corrected with a medial canthopexy, which has been shown to have better outcomes than the lateral counterpart. [155] Reanimation of eyelid closure can be accomplished with nerve or muscle transfers. Cross-facial nerve grafting, hypoglossal nerve transfer, or direct orbicularis oculi neurotization can restore innervation to the orbicularis oculi muscle. [156] Alternatively, muscle transposition from the frontalis or temporalis and muscle transfers, commonly from the gracilis, can be used to achieve eyelid closure. [156, 157] In the case of muscle transfers, a cross-nerve graft from the contralateral side reinnervates the fasciscular territory. Nerve transfers achieve better eyelid closure and blinking than muscle transfers, however. [156]

Form and Function Considerations of Facial Paresis

Injury to the occipitofrontalis muscle can lead to brow ptosis, which leads to cosmetic disfigurement and drooping into the upper eye, exacerbating dermatochalasis and obstructs the upper visual field. Surgical brow elevation can be done with open brow approaches or endoscopically, with similar results. [158]

Paralysis of the midface poses a concern because it can exacerbate lower lid ectropion, by exerting weight on the canthal ligaments, and cause nasal valve stenosis. The extended minimal access cranial suspension lift can bolster the support on the lower lid, by lifting the midface. [159]

Symptomatic stenosis of the nasal valve can be surgically corrected by suspending sutures to fix the nasal valve to the sidewall. While this provides short term relief of the obstruction, it typically reverses with time. A more extensive suture suspension involving the nose, lateral edge of the mouth and the lateral canthus has been described to provide relief of obstruction in the long-term. [160] Percutaneous Gore-tex sling to suspend the nasolabial fold has also shown to provide benefit. [161] In addition to improving mid-face symmetry, rhytidectomy has been shown to open the nasal canal as well. [162] Lastly, functional septorhinoplasty may be beneficial for alleviating nasal obstruction in patients with facial palsy.

Facial paralysis in the lower face causes ptyalism, difficulty with eating and drinking, poor articulation, and loss of smile and lower face expressivity. At rest, loss of muscle tension in the levator labii superioris, zygomaticus major and minor causes drooping of the lateral mouth. Static slings from the zygomatic arch or deep temporal fascia can be used to elevate the upper lip and recreate the nasolabial fold. [163] Facial lifting procedures can also be used to statically elevate the lower face. Lip reanimation can be accomplished by transferring the palmaris longus tendon. [164]

NONOPERATIVE MANAGEMENT

Botulinum toxin A works well for treating facial synkinesis, and it has been shown to improve patients' quality of life. [165] In the event of surgical failure, botulinum toxin can also be injected into the non-paralyzed side of the face to improve facial symmetry. [166]

PHYSICAL THERAPY

In facial reanimation therapy, patients work with a therapist to retrain their muscles facial expression. Therapists educate patients about rudimentary facial anatomy and function, which helps patients isolate and control muscles that are normally under involuntary control. Using a variety of visual, proprioceptive, and sensory feedback, patients train their muscles between therapy sessions. In particular, surface EMG, which detects muscle activity from electrodes placed on a patient's skin, has been shown to be effective in improving the degree of facial paralysis and synkinesis. [167] In addition, mirror biofeedback therapy has been shown to reduce ocular synkinesis during retraining muscles involved with smiling. While it cannot completely eliminate synkinesis, most patients experience significant improvements in both the extent of voluntary movements and level of facial symmetry. Neuromuscular retraining therapy can be completed between 18 months and three years, with most patients practicing the therapy at home. [168] Patients who are cognitively intact and are self-motivated will benefit. Even children as young as 8 have shown improvement after neuromuscular retraining. [169]

Patients with synkinesis have an increased resting muscle tone which can limit their functional range of motion and cause pain. This increased tone is commonly seen in the nasolabial folds or platysma. Massage, soft tissue mobilization, and heat compresses can be used to reduce muscle tone. [170]

Mime therapy combines facial reanimation techniques with facial expression exercises. The goal of mime therapy is to integrate conscious facial movements with emotional expressions. It also includes massages and stretching exercises. Mime therapy has been shown to produce lasting improvements in facial expression, graded on the Sunnybrook and House-Brackmann scales. [171–173]

Brief electrical stimulation of the facial nerve has been investigated to improve recovery, based on the assumption that the injured nerve begins to atrophy due to lack of stimulation. However, patients with synkinesis have an increased muscle tone, possibly due to hyperstimulation. In practice, brief electrical stimulation has not been shown to have a benefit and may even exacerbate paralysis.

REFERENCES

[1] Kochhar A, Larian B, Azizzadeh B. Facial Nerve and Parotid Gland Anatomy. *Otolaryngol Clin North Am.* 2016;49(2):273-284. doi:10.1016/j.otc.2015.10.002.

[2] Moore GF. Facial nerve paralysis. *Prim Care.* 1990;17(2):437-460.

[3] Kikuta S, Iwanaga J, Watanabe K, Kusukawa J, Tubbs RS. The Feasibility of Using the Posterior Auricular Branch of the Facial Nerve as a Donor for Facial Nerve Reanimation Procedures: A Cadaveric Study. *J Oral Maxillofac Surg.* 2019;77(7):1470.e1-1470.e8. doi:10.1016/j.joms.2019.02.043.

[4] Trainor PA, Krumlauf R. Patterning the cranial neural crest: hindbrain segmentation and Hox gene plasticity. *Nat Rev Neurosci.* 2000;1(2):116-124. doi:10.1038/35039056.

[5] Coulson SE, O'dwyer NJ, Adams RD, Croxson GR. Expression of emotion and quality of life after facial nerve paralysis. *Otol Neurotol Off Publ Am Otol Soc Am Neurotol Soc Eur Acad Otol Neurotol.* 2004;25(6):1014-1019. doi:10.1097/00129492-200411000-00026.

[6] Schulkin J. *The Neuroendocrine Regulation of Behavior.* Cambridge University Press; 1998. doi:10.1017/CBO9780511818738.

[7] Hains SMJ, Muir DW. Infant Sensitivity to Adult Eye Direction. *Child Dev.* 1996;67(5):1940-1951. doi:10.1111/j.1467-8624.1996.tb01836.x.

[8] Knappmeyer B, Thornton IM, Bülthoff HH. The use of facial motion and facial form during the processing of identity. *Vision Res.* 2003;43(18):1921-1936. doi:10.1016/S0042-6989(03)00236-0.

[9] Erickson K, Schulkin J. Facial expressions of emotion: a cognitive neuroscience perspective. *Brain Cogn.* 2003;52(1):52-60. doi:10.1016/s0278-2626(03)00008-3.

[10] VanSwearingen JM, Cohn JF, Turnbull J, Mrzai T, Johnson P. Psychological distress: linking impairment with disability in facial neuromotor disorders. *Otolaryngol--Head Neck Surg Off J Am Acad Otolaryngol-Head Neck Surg.* 1998;118(6):790-796. doi:10.1016/S0194-5998(98)70270-0.

[11] Ross BG, Fradet G, Nedzelski JM. Development of a sensitive clinical facial grading system. *Otolaryngol--Head Neck Surg Off J Am Acad Otolaryngol-Head Neck Surg.* 1996;114(3):380-386. doi:10.1016/s0194-5998(96)70206-1.

[12] Fattah AY, Gurusinghe ADR, Gavilan J, et al. Facial nerve grading instruments: systematic review of the literature and suggestion for uniformity. *Plast Reconstr Surg.* 2015;135(2):569-579. doi:10.1097/PRS.0000000000000905.

[13] Kahn JB, Gliklich RE, Boyev KP, Stewart MG, Metson RB, McKenna MJ. Validation of a patient-graded instrument for facial nerve paralysis: the FaCE scale. *The Laryngoscope.* 2001;111(3):387-398. doi:10.1097/00005537-200103000-00005.

[14] Mehta RP, WernickRobinson M, Hadlock TA. Validation of the Synkinesis Assessment Questionnaire. *The Laryngoscope.* 2007;117(5):923-926. doi:10.1097/MLG.0b013e3180412460.

[15] VanSwearingen JM, Brach JS. The Facial Disability Index: reliability and validity of a disability assessment instrument for disorders of the facial neuromuscular system. *Phys Ther.* 1996;76(12):1288-1298; discussion 1298-1300. doi:10.1093/ptj/76.12.1288.

[16] Goodridge AE. Electromyography and Nerve Conduction Studies. *Can Fam Physician.* 1988;34:339-343.

[17] Gupta S, Mends F, Hagiwara M, Fatterpekar G, Roehm PC. Imaging the Facial Nerve: A Contemporary Review. *Radiol Res Pract.* 2013;2013. doi:10.1155/2013/248039.

[18] Wilson DF, Talbot JM, Hodgson RS. Magnetic Resonance Imaging–Enhancing Lesions of the Labyrinth and Facial Nerve: Clinical Correlation. *Arch Otolaryngol Neck Surg.* 1994;120(5):560-564. doi:10.1001/archotol.1994.01880290070012.

[19] Hilly O, Chen JM, Birch J, et al. Diffusion Tensor Imaging Tractography of the Facial Nerve in Patients With Cerebellopontine Angle Tumors. *Otol Neurotol Off Publ Am Otol Soc Am Neurotol Soc Eur Acad Otol Neurotol.* 2016;37(4):388-393. doi:10.1097/MAO.0000000000000984.

[20] Rouchy R-C, Attyé A, Troprès I, et al. Facial nerve tractography: A new tool to detect perineural invasion in parotid cancers. *J Neuroradiol.* 2017;44(2):74-75. doi:10.1016/j.neurad.2017.01.005.

[21] Chu J, Zhou Z, Hong G, et al. High-resolution MRI of the intraparotid facial nerve based on a microsurface coil and a 3D reversed fast imaging with steady-state precession DWI sequence at 3T. *AJNR Am J Neuroradiol.* 2013;34(8):1643-1648. doi:10.3174/ajnr.A3472.

[22] Gebarski SS, Telian SA, Niparko JK. Enhancement along the normal facial nerve in the facial canal: MR imaging and anatomic correlation. *Radiology.* 1992;183(2):391-394. doi:10.1148/radiology.183.2.1561339.

[23] May M, Klein SR. Differential diagnosis of facial nerve palsy. *Otolaryngol Clin North Am.* 1991;24(3):613-645.

[24] Gondivkar S, Parikh V, Parikh R. Herpes zoster oticus: A rare clinical entity. *Contemp Clin Dent.* 2010;1(2):127-129. doi:10.4103/0976-237X.68588.

[25] Aviel A, Ostfeld E, Burstein R, Marshak G, Bentwich Z. Peripheral blood T and B lymphocyte subpopulations in Bell's palsy. *Ann Otol*

Rhinol Laryngol. 1983;92(2 Pt 1):187-191. doi:10.1177/0003 48948309200218.

[26] Greco A, Gallo A, Fusconi M, Marinelli C, Macri GF, de Vincentiis M. Bell's palsy and autoimmunity. *Autoimmun Rev.* 2012;12(2):323-328. doi:10.1016/j.autrev.2012.05.008.

[27] Schirm J, Mulkens PS. Bell's palsy and herpes simplex virus. *APMIS Acta Pathol Microbiol Immunol Scand.* 1997;105(11):815-823. doi:10.1111/j.1699-0463.1997.tb05089.x.

[28] Baringer JR. Herpes simplex virus and Bell palsy. *Ann Intern Med.* 1996;124(1 Pt 1):63-65. doi:10.7326/0003-4819-124-1_part_1-199601010-00010.

[29] Peitersen E. Bell's palsy: the spontaneous course of 2,500 peripheral facial nerve palsies of different etiologies. *Acta Oto-Laryngol Suppl.* 2002;(549):4-30.

[30] Mutsch M, Zhou W, Rhodes P, et al. Use of the inactivated intranasal influenza vaccine and the risk of Bell's palsy in Switzerland. *N Engl J Med.* 2004;350(9):896-903. doi:10.1056/NEJMoa030595.

[31] Holland NJ, Weiner GM. Recent developments in Bell's palsy. *BMJ.* 2004;329(7465):553-557.

[32] Bodénez C, Bernat I, Willer J-C, Barré P, Lamas G, Tankéré F. Facial nerve decompression for idiopathic Bell's palsy: report of 13 cases and literature review. *J Laryngol Otol.* 2010;124(3):272-278. doi:10.1017/S0022215109991265.

[33] Benecke JE. Facial paralysis. *Otolaryngol Clin North Am.* 2002;35(2):357-365. doi:10.1016/s0030-6665(02)00003-8.

[34] Wolfson AB, Cloutier R, Hendey GW, Ling LJ, Rosen CL, Schaider JJ. *Harwood-Nuss' Clinical Practice of Emergency Medicine: Sixth Edition.* Wolters Kluwer Health Adis (ESP); 2014. Accessed September 5, 2020. https://ohsu.pure.elsevier.com/en/publications/harwood-nuss-clinical-practice-of-emergency-medicine-sixth-editio.

[35] Adour KK, Ruboyianes JM, Von Doersten PG, et al. Bell's palsy treatment with acyclovir and prednisone compared with prednisone

alone: a double-blind, randomized, controlled trial. *Ann Otol Rhinol Laryngol.* 1996;105(5):371-378. doi:10.1177/000348949610500508.

[36] Browning GG. Bell's palsy: a review of three systematic reviews of steroid and anti-viral therapy. *Clin Otolaryngol.* 2010;35(1):56-58. doi:10.1111/j.1749-4486.2010.02084.x.

[37] de Almeida JR, Al Khabori M, Guyatt GH, et al. Combined corticosteroid and antiviral treatment for Bell palsy: a systematic review and meta-analysis. *JAMA.* 2009;302(9):985-993. doi:10.1001/jama.2009.1243.

[38] Vespa PM, Nuwer MR, Nenov V, et al. Increased incidence and impact of nonconvulsive and convulsive seizures after traumatic brain injury as detected by continuous electroencephalographic monitoring. *J Neurosurg.* 1999;91(5):750-760. doi:10.3171/jns.1999.91.5.0750.

[39] Hato N, Yamada H, Kohno H, et al. Valacyclovir and prednisolone treatment for Bell's palsy: a multicenter, randomized, placebo-controlled study. *Otol Neurotol Off Publ Am Otol Soc Am Neurotol Soc Eur Acad Otol Neurotol.* 2007;28(3):408-413. doi:10.1097/01.mao.0000265190.29969.12.

[40] Sullivan FM, Swan IRC, Donnan PT, et al. Early Treatment with Prednisolone or Acyclovir in Bell's Palsy. *N Engl J Med.* 2007;357(16):1598-1607. doi:10.1056/NEJMoa072006.

[41] Numthavaj P, Thakkinstian A, Dejthevaporn C, Attia J. Corticosteroid and antiviral therapy for Bell's palsy: A network meta-analysis. *BMC Neurol.* 2011;11(1):1. doi:10.1186/1471-2377-11-1.

[42] Yeo SG, Lee YC, Park DC, Cha CI. Acyclovir plus steroid vs steroid alone in the treatment of Bell's palsy. *Am J Otolaryngol.* 2008;29(3):163-166. doi:10.1016/j.amjoto.2007.05.001.

[43] Quant EC, Jeste SS, Muni RH, Cape AV, Bhussar MK, Peleg AY. The benefits of steroids versus steroids plus antivirals for treatment of Bell's palsy: a meta-analysis. *BMJ.* 2009;339:b3354. doi:10.1136/bmj.b3354.

[44] Lee HY, Byun JY, Park MS, Yeo SG. Steroid-antiviral treatment improves the recovery rate in patients with severe Bell's palsy. *Am J Med.* 2013;126(4):336-341. doi:10.1016/j.amjmed.2012.08.020.

[45] Gantz BJ, Rubinstein JT, Gidley P, Woodworth GG. Surgical management of Bell's palsy. *The Laryngoscope.* 1999;109(8):1177-1188. doi:10.1097/00005537-199908000-00001.

[46] Chen N, Zhou M, He L, Zhou D, Li N. Acupuncture for Bell's palsy. *Cochrane Database Syst Rev.* 2010;(8):CD002914. doi:10.1002/14651858.CD002914.pub5.

[47] Chang CY, Cass SP. Management of facial nerve injury due to temporal bone trauma. *Am J Otol.* 1999;20(1):96-114.

[48] Brodie HA, Thompson TC. Management of complications from 820 temporal bone fractures. *Am J Otol.* 1997;18(2):188-197.

[49] McKennan KX, Chole RA. Facial paralysis in temporal bone trauma. *Am J Otol.* 1992;13(2):167-172.

[50] Aguilar EA, Yeakley JW, Ghorayeb BY, Hauser M, Cabrera J, Jahrsdoerfer RA. High resolution CT scan of temporal bone fractures: association of facial nerve paralysis with temporal bone fractures. *Head Neck Surg.* 1987;9(3):162-166. doi:10.1002/hed.2890090306.

[51] Little SC, Kesser BW. Radiographic Classification of Temporal Bone Fractures: Clinical Predictability Using a New System. *Arch Otolaryngol Neck Surg.* 2006;132(12):1300-1304. doi:10.1001/archotol.132.12.1300.

[52] Coker NJ, Kendall KA, Jenkins HA, Alford BR. Traumatic intratemporal facial nerve injury: management rationale for preservation of function. *Otolaryngol--Head Neck Surg Off J Am Acad Otolaryngol-Head Neck Surg.* 1987;97(3):262-269. doi:10.1177/019459988709700303.

[53] Terzis JK, Konofaos P. Nerve transfers in facial palsy. *Facial Plast Surg FPS.* 2008;24(2):177-193. doi:10.1055/s-2008-1075833.

[54] Dahiya R, Keller JD, Litofsky NS, Bankey PE, Bonassar LJ, Megerian CA. Temporal bone fractures: otic capsule sparing versus otic capsule violating clinical and radiographic considerations. *J*

Trauma. 1999;47(6):1079-1083. doi:10.1097/00005373-199912000-00014.

[55] Wayman DM, Pham HN, Byl FM, Adour KK. Audiological manifestations of Ramsay Hunt syndrome. *J Laryngol Otol.* 1990;104(2):104-108. doi:10.1017/s0022215100111971.

[56] Hato N, Kisaki H, Honda N, Gyo K, Murakami S, Yanagihara N. Ramsay Hunt syndrome in children. *Ann Neurol.* 2000;48(2):254-256.

[57] Devriese PP, Moesker WH. The natural history of facial paralysis in herpes zoster. *Clin Otolaryngol Allied Sci.* 1988;13(4):289-298. doi:10.1111/j.1365-2273.1988.tb01134.x.

[58] Morgan M, Nathwani D. Facial palsy and infection: the unfolding story. *Clin Infect Dis Off Publ Infect Dis Soc Am.* 1992;14(1):263-271. doi:10.1093/clinids/14.1.263.

[59] Terada K, Niizuma T, Kawano S, Kataoka N, Akisada T, Orita Y. Detection of varicella-zoster virus DNA in peripheral mononuclear cells from patients with Ramsay Hunt syndrome or zoster sine herpete. *J Med Virol.* 1998;56(4):359-363.

[60] Hunt JR. The symptom-complex of the acute posterior poliomyelitis of the geniculate, auditory, glossopharyngeal and pneumogastric ganglia. *Arch Intern Med.* 1910;V(6):631-675. doi:10.1001/archinte.1910.00050280097006.

[61] Robillard RB, Hilsinger RL, Adour KK. Ramsay Hunt facial paralysis: clinical analyses of 185 patients. *Otolaryngol--Head Neck Surg Off J Am Acad Otolaryngol-Head Neck Surg.* 1986;95(3 Pt 1):292-297. doi:10.1177/01945998860953P105.

[62] Aleksic SN, Budzilovich GN, Lieberman AN. Herpes zoster oticus and facial paralysis (Ramsay Hunt syndrome). Clinico-pathologic study and review of literature. *J Neurol Sci.* 1973;20(2):149-159. doi:10.1016/0022-510x(73)90027-0.

[63] Denny-Brown D, Adams RD, Fitzgerald PJ. Pathologic features of herpes zoster: a note on geniculate herpes. *Arch Neurol Psychiatry.* 1944;51(3):216-231. doi:10.1001/archneurpsyc.1944.02290270005002.

[64] B B, I F, I W. Herpes zoster auris associated with facial nerve palsy and auditory nerve symptoms: a case report with histopathological findings. *Acta Otolaryngol (Stockh)*. 1967;63(6):533-550. doi:10.3109/00016486709128786.

[65] Guldberg-Möller J, Olsen S, Kettel K. Histopathology of the Facial Nerve in Herpes Zoster Oticus. *AMA Arch Otolaryngol*. 1959;69(3):266-275. doi:10.1001/archotol.1959.00730030274003.

[66] Murakami S, Honda N, Mizobuchi M, Nakashiro Y, Hato N, Gyo K. Rapid diagnosis of varicella zoster virus infection in acute facial palsy. *Neurology*. 1998;51(4):1202-1205. doi:10.1212/wnl.51.4.1202.

[67] Jonsson L, Tien R, Engström M, Thuomas KA. Gd-DPTA enhanced MRI in Bell's palsy and herpes zoster oticus: an overview and implications for future studies. *Acta Otolaryngol (Stockh)*. 1995;115(5):577-584. doi:10.3109/00016489509139371.

[68] Murakami S, Hato N, Horiuchi J, Honda N, Gyo K, Yanagihara N. Treatment of Ramsay Hunt syndrome with acyclovir-prednisone: significance of early diagnosis and treatment. *Ann Neurol*. 1997;41(3):353-357. doi:10.1002/ana.410410310.

[69] Clark JR, Carlson RD, Sasaki CT, Pachner AR, Steere AC. Facial paralysis in Lyme disease. *The Laryngoscope*. 1985;95(11):1341-1345.

[70] Kalish RA, Kaplan RF, Taylor E, Jones-Woodward L, Workman K, Steere AC. Evaluation of study patients with Lyme disease, 10-20-year follow-up. *J Infect Dis*. 2001;183(3):453-460. doi:10.1086/318082.

[71] Halperin JJ. Nervous system Lyme disease. *Infect Dis Clin North Am*. 2015;29(2):241-253. doi:10.1016/j.idc.2015.02.002.

[72] Rauer S, Kastenbauer S, Fingerle V, Hunfeld K-P, Huppertz H-I, Dersch R. Lyme Neuroborreliosis. *Dtsch Ärztebl Int*. 2018;115(45):751-756. doi:10.3238/arztebl.2018.0751.

[73] Halperin JJ, Shapiro ED, Logigian E, et al. Practice parameter: treatment of nervous system Lyme disease (an evidence-based review): report of the Quality Standards Subcommittee of the

American Academy of Neurology. *Neurology.* 2007;69(1):91-102. doi:10.1212/01.wnl.0000265517.66976.28.

[74] Jowett N, Gaudin RA, Banks CA, Hadlock TA. Steroid use in Lyme disease-associated facial palsy is associated with worse long-term outcomes. *The Laryngoscope.* 2017;127(6):1451-1458. doi:10.1002/lary.26273.

[75] Jowett N, Hadlock TA. A Contemporary Approach to Facial Reanimation. *JAMA Facial Plast Surg.* 2015;17(4):293-300. doi:10.1001/jamafacial.2015.0399.

[76] Samuel J, Fernandes CMC. Tuberculous Mastoiditis: *Ann Otol Rhinol Laryngol.* Published online June 29, 2016. doi:10.1177/000348948609500310.

[77] M'Cart HWD. Tuberculous Disease of the Middle Ear. *J Laryngol Otol.* 1925;40(7):456-466. doi:10.1017/S0022215100027626.

[78] Pallen MJ, Butcher PD. New strategies in microbiological diagnosis. *J Hosp Infect.* 1991;18 Suppl A:147-158. doi:10.1016/0195-6701(91)90017-3.

[79] Singh B. Role of surgery in tuberculous mastoiditis. *J Laryngol Otol.* 1991;105(11):907-915. doi:10.1017/s0022215100117797.

[80] Bluestone CD. Otitis media, atelectasis, and eustachian tube dysfunction. *Pediatr Otolaryngol.* Published online 1990:320-486.

[81] Harbert F, Riordan D. Tuberculosis of the middle ear. *The Laryngoscope.* 1964;74:198-204. doi:10.1002/lary.5540740203.

[82] Myerson MC, Gilbert JG. Tuberculosis of the middle ear and mastoid. *Arch Otolaryngol.* 1941;33(2):231-250. doi:10.1001/archotol.1941.00660030234007.

[83] Lipkin WI, Parry G, Kiprov D, Abrams D. Inflammatory neuropathy in homosexual men with lymphadenopathy. *Neurology.* 1985;35(10):1479-1483. doi:10.1212/wnl.35.10.1479.

[84] Miller RG, Parry GJ, Pfaeffl W, Lang W, Lippert R, Kiprov D. The spectrum of peripheral neuropathy associated with ARC and AIDS. *Muscle Nerve.* 1988;11(8):857-863. doi:10.1002/mus.880110810.

[85] de la Monte SM, Gabuzda DH, Ho DD, et al. Peripheral neuropathy in the acquired immunodeficiency syndrome. *Ann Neurol.* 1988;23(5):485-492. doi:10.1002/ana.410230510.

[86] Cornblath DR, McArthur JC, Kennedy PG, Witte AS, Griffin JW. Inflammatory demyelinating peripheral neuropathies associated with human T-cell lymphotropic virus type III infection. *Ann Neurol.* 1987;21(1):32-40. doi:10.1002/ana.410210107.

[87] Dalakas MC, Pezeshkpour GH. Neuromuscular diseases associated with human immunodeficiency virus infection. *Ann Neurol.* 1988;23 Suppl:S38-48. doi:10.1002/ana.410230713.

[88] Belec L, Georges AJ, Bouree P, et al. Peripheral facial nerve palsy related to HIV infection: relationship with the immunological status and the HIV staging in Central Africa. *Cent Afr J Med.* 1991;37(3):88-93.

[89] Brown MM, Thompson A, Goh BT, Forster GE, Swash M. Bell's palsy and HIV infection. *J Neurol Neurosurg Psychiatry.* 1988;51(3):425-426.

[90] Snider WD, Simpson DM, Nielsen S, Gold JW, Metroka CE, Posner JB. Neurological complications of acquired immune deficiency syndrome: analysis of 50 patients. *Ann Neurol.* 1983;14(4):403-418. doi:10.1002/ana.410140404.

[91] Gray F, Gherardi R, Scaravilli F. The Neuropathology of the Acquired Immune Deficiency Syndrome (Aids)A Review. *Brain.* 1988;111(2):245-266. doi:10.1093/brain/111.2.245.

[92] Moses PD, Pereira SM, John TJ, Steinhoff M. Poliovirus infection and Bell's palsy in children. *Ann Trop Paediatr.* 1985;5(4):195-196. doi:10.1080/02724936.1985.11748391.

[93] Endo A, Izumi H, Miyashita M, Okubo O, Harada K. Facial palsy associated with mumps parotitis. *Pediatr Infect Dis J.* 2001; 20(8):815–816.

[94] Johnson PA, Avery C. Infectious mononucleosis presenting as a parotid mass with associated facial nerve palsy. *Int J Oral Maxillofac Surg.* 1991;20(4):193-195. doi:10.1016/s0901-5027 (05)80171-7.

[95] Lubbers WJ, Schipper A, Hogeweg M, de Soldenhoff R. Paralysis of facial muscles in leprosy patients with lagophthalmos. *Int J Lepr Mycobact Dis Off Organ Int Lepr Assoc*. 1994;62(2):220-224.

[96] Chiu AG, Hecht DA, Prendiville SA, Mesick C, Mikula S, Deeb ZE. Atypical presentations of cat scratch disease in the head and neck. *Otolaryngol--Head Neck Surg Off J Am Acad Otolaryngol-Head Neck Surg*. 2001;125(4):414-416. doi:10.1067/mhn.2001.116792.

[97] Patey O, Ollivaud L, Breuil J, Lafaix C. Unusual neurologic manifestations occurring during dengue fever infection. *Am J Trop Med Hyg*. 1993;48(6):793-802. doi:10.4269/ajtmh.1993.48.793.

[98] Fukazawa T, Tashiro K, Hamada T, et al. Multiple Sclerosis in Hokkaido, the Northernmost Island of Japan: Prospective Analyses of Clinical Features. *Intern Med*. 1992;31(3):349-352. doi:10.2169/internalmedicine.31.349.

[99] Hung TP, Landsborough D, Hsi MS. Multiple sclerosis amongst Chinese in Taiwan. *J Neurol Sci*. 1976;27(4):459-484. doi:10.1016/0022-510x(76)90214-8.

[100] Kurtzke JF, Beebe GW, Nagler B, Auth TL, Kurland LT, Nefzger MD. Studies on natural history of multiple sclerosis. 4. Clinical features of the onset bout. *Acta Neurol Scand*. 1968;44(4):467-494. doi:10.1111/j.1600-0404.1968.tb05587.x.

[101] Carter S, Sciarra D, Merritt HH. The course of multiple sclerosis as determined by autopsy proven cases. *Res Publ - Assoc Res Nerv Ment Dis*. 1950;28:471-511.

[102] Kwon JY, Kim JY, Jeong JH, Park KD. Multiple sclerosis and peripheral multifocal demyelinating neuropathies occurring in a same patient. *J Clin Neurol Seoul Korea*. 2008;4(1):51-57. doi:10.3988/jcn.2008.4.1.51.

[103] Golden SK, Reiff CJ, Painter CJ, Repplinger MD. Myasthenia Gravis Presenting as Persistent Unilateral Ptosis with Facial Droop. *J Emerg Med*. 2015;49(1):e23-e25. doi:10.1016/j.jemermed.2015.01.002.

[104] Narayanan RP, James N, Ramachandran K, Jaramillo MJ. Guillain-Barré Syndrome presenting with bilateral facial nerve paralysis: a case report. *Cases J.* 2008;1:379. doi:10.1186/1757-1626-1-379.

[105] Sharma K, Tengsupakul S, Sanchez O, Phaltas R, Maertens P. Guillain–Barré syndrome with unilateral peripheral facial and bulbar palsy in a child: A case report. *SAGE Open Med Case Rep.* 2019;7. doi:10.1177/2050313X19838750.

[106] Keane JR. Bilateral seventh nerve palsy: analysis of 43 cases and review of the literature. *Neurology.* 1994;44(7):1198-1202. doi:10.1212/wnl.44.7.1198.

[107] Zuccoli G, Panigrahy A, Bailey A, Fitz C. Redefining the Guillain-Barré spectrum in children: neuroimaging findings of cranial nerve involvement. *AJNR Am J Neuroradiol.* 2011;32(4):639-642. doi:10.3174/ajnr.A2358.

[108] Mori M, Kuwabara S, Fukutake T, Yuki N, Hattori T. Clinical features and prognosis of Miller Fisher syndrome. *Neurology.* 2001;56(8):1104-1106. doi:10.1212/wnl.56.8.1104.

[109] Ozgursoy OB, Karatayli Ozgursoy S, Tulunay O, Kemal O, Akyol A, Dursun G. Melkersson-Rosenthal syndrome revisited as a misdiagnosed disease. *Am J Otolaryngol.* 2009;30(1):33-37. doi:10.1016/j.amjoto.2008.02.004.

[110] Gerressen M, Ghassemi A, Stockbrink G, Riediger D, Zadeh MD. Melkersson-Rosenthal Syndrome: Case Report of a 30-Year Misdiagnosis. *J Oral Maxillofac Surg.* 2005;63(7):1035-1039. doi:10.1016/j.joms.2005.03.021.

[111] Kanerva M, Moilanen K, Virolainen S, Vaheri A, Pitkäranta A. Melkersson-Rosenthal syndrome. *Otolaryngol Neck Surg.* 2008;138(2):246-251. doi:10.1016/j.otohns.2007.11.015.

[112] Greene RM, Rogers RS. Melkersson-Rosenthal syndrome: a review of 36 patients. *J Am Acad Dermatol.* 1989;21(6):1263-1270. doi:10.1016/s0190-9622(89)70341-8.

[113] Bruns AD, Burgess LP. Familial recurrent facial paresis: four generations. *Otolaryngol--Head Neck Surg Off J Am Acad*

Otolaryngol-Head Neck Surg. 1998;118(6):859-862. doi:10.1016/ S0194-5998(98)70283-9.

[114] Shapiro M, Peters S, Spinelli HM. Melkersson-Rosenthal syndrome in the periocular area: a review of the literature and case report. *Ann Plast Surg.* 2003;50(6):644-648. doi:10.1097/01.SAP.0000069068. 03742.48.

[115] Wehl G, Rauchenzauner M. A Systematic Review of the Literature of the Three Related Disease Entities Cheilitis Granulomatosa, Orofacial Granulomatosis and Melkersson - Rosenthal Syndrome. *Curr Pediatr Rev.* 2018;14(3):196-203. doi:10.2174/157339631 4666180515113941.

[116] Graham MD, Kartush JM. Total facial nerve decompression for recurrent facial paralysis: an update. *Otolaryngol--Head Neck Surg Off J Am Acad Otolaryngol-Head Neck Surg.* 1989;101(4):442-444. doi:10.1177/019459988910100406.

[117] Graham MD, Kemink JL. Total facial nerve decompression in recurrent facial paralysis and the Melkersson-Rosenthal syndrome: a preliminary report. *Am J Otol.* 1986;7(1):34-37.

[118] Picciolini O, Porro M, Cattaneo E, et al. Moebius syndrome: clinical features, diagnosis, management and early intervention. *Ital J Pediatr.* 2016;42(1):56. doi:10.1186/s13052-016-0256-5.

[119] Stern BJ, Krumholz A, Johns C, Scott P, Nissim J. Sarcoidosis and its neurological manifestations. *Arch Neurol.* 1985;42(9):909-917. doi:10.1001/archneur.1985.04060080095022.

[120] Zajicek JP, Scolding NJ, Foster O, et al. Central nervous system sarcoidosis—diagnosis and management. *QJM Int J Med.* 1999;92(2):103-117. doi:10.1093/qjmed/92.2.103.

[121] Christoforidis GA, Spickler EM, Recio MV, Mehta BM. MR of CNS sarcoidosis: correlation of imaging features to clinical symptoms and response to treatment. *AJNR Am J Neuroradiol.* 1999;20(4):655-669.

[122] Gaines JD, Eckman PB, Remington JS. Low CSF glucose level in sarcoidosis involving the central nervous system. *Arch Intern Med.* 1970;125(2):333-336.

[123] H S, R C, E O, C A. The challenge of profound hypoglycorrhachia: two cases of sarcoidosis and review of the literature. *Clin Rheumatol.* 2011;30(12):1631-1639. doi:10.1007/s10067-011-1834-y.

[124] Joseph FG, Scolding NJ. Neurosarcoidosis: a study of 30 new cases. *J Neurol Neurosurg Psychiatry.* 2009;80(3):297-304. doi:10.1136/jnnp.2008.151977.

[125] Riga M, Kefalidis G, Danielides V. The Role of Diabetes Mellitus in the Clinical Presentation and Prognosis of Bell Palsy. *J Am Board Fam Med.* 2012;25(6):819-826. doi:10.3122/jabfm.2012.06.120084.

[126] Teller DC, Murphy TP. Bilateral facial paralysis: a case presentation and literature review. *J Otolaryngol.* 1992;21(1):44-47.

[127] Sunderland IRP, Brenner MJ, Singham J, Rickman SR, Hunter DA, Mackinnon SE. Effect of tension on nerve regeneration in rat sciatic nerve transection model. *Ann Plast Surg.* 2004;53(4):382-387. doi:10.1097/01.sap.0000125502.63302.47.

[128] Myckatyn TM, Mackinnon SE. The surgical management of facial nerve injury. *Clin Plast Surg.* 2003;30(2):307-318. doi:10.1016/S0094-1298(02)00102-5.

[129] Samii M, Matthies C. Indication, technique and results of facial nerve reconstruction. *Acta Neurochir (Wien).* 1994;130(1):125-139. doi:10.1007/BF01405512.

[130] Volk GF, Pantel M, Streppel M, Guntinas-Lichius O. Reconstruction of complex peripheral facial nerve defects by a combined approach using facial nerve interpositional graft and hypoglossal-facial jump nerve suture. *The Laryngoscope.* 2011;121(11):2402-2405. doi:10.1002/lary.22357.

[131] Scaramella LF. Anastomosis between the two facial nerves. *The Laryngoscope.* 1975;85(8):1359-1366. doi:10.1288/00005537-197508000-00012.

[132] Lee EI, Hurvitz KA, Evans GRD, Wirth GA. Cross-facial nerve graft: past and present. *J Plast Reconstr Aesthetic Surg JPRAS.* 2008;61(3):250-256. doi:10.1016/j.bjps.2007.05.016.

[133] Galli SKD, Valauri F, Komisar A. Facial reanimation by cross-facial nerve grafting: report of five cases. *Ear Nose Throat J.* 2002;81(1):25-29.

[134] Gousheh J, Arasteh E. Treatment of facial paralysis: dynamic reanimation of spontaneous facial expression-apropos of 655 patients. *Plast Reconstr Surg.* 2011;128(6):693e-703e. doi:10.1097/PRS.0b013e318230c58f.

[135] Conley J, Baker DC. Hypoglossal-facial nerve anastomosis for reinnervation of the paralyzed face. *Plast Reconstr Surg.* 1979;63(1):63-72. doi:10.1097/00006534-197901000-00011.

[136] Rochkind S, Shafi M, Alon M, Salame K, Fliss DM. Facial nerve reconstruction using a split hypoglossal nerve with preservation of tongue function. *J Reconstr Microsurg.* 2008;24(7):469-474. doi:10.1055/s-0028-1088225.

[137] Arai H, Sato K, Yanai A. Hemihypoglossal-facial nerve anastomosis in treating unilateral facial palsy after acoustic neurinoma resection. *J Neurosurg.* 1995;82(1):51-54. doi:10.3171/jns.1995.82.1.0051.

[138] Hayashi A, Nishida M, Seno H, et al. Hemihypoglossal nerve transfer for acute facial paralysis: Clinical article. *J Neurosurg.* 2013;118(1):160-166. doi:10.3171/2012.9.JNS1270.

[139] Corrales CE, Gurgel RK, Jackler RK. Rehabilitation of central facial paralysis with hypoglossal-facial anastomosis. *Otol Neurotol Off Publ Am Otol Soc Am Neurotol Soc Eur Acad Otol Neurotol.* 2012;33(8):1439-1444. doi:10.1097/MAO.0b013e3182693cd0.

[140] Hontanilla B, Marré D. Comparison of hemihypoglossal nerve versus masseteric nerve transpositions in the rehabilitation of short-term facial paralysis using the Facial Clima evaluating system. *Plast Reconstr Surg.* 2012;130(5):662e-672e. doi:10.1097/PRS.0b013e318267d5e8.

[141] Schaverien M, Moran G, Stewart K, Addison P. Activation of the masseter muscle during normal smile production and the implications for dynamic reanimation surgery for facial paralysis. *J Plast Reconstr Aesthetic Surg JPRAS.* 2011;64(12):1585-1588. doi:10.1016/j.bjps.2011.07.012.

[142] Faria JCM, Scopel GP, Busnardo FF, Ferreira MC. Nerve sources for facial reanimation with muscle transplant in patients with unilateral facial palsy: clinical analysis of 3 techniques. *Ann Plast Surg*. 2007;59(1):87-91. doi:10.1097/01.sap.0000252042.58200.c3.

[143] Borschel GH, Kawamura DH, Kasukurthi R, Hunter DA, Zuker RM, Woo AS. The motor nerve to the masseter muscle: an anatomic and histomorphometric study to facilitate its use in facial reanimation. *J Plast Reconstr Aesthetic Surg JPRAS*. 2012;65(3):363-366. doi:10.1016/j.bjps.2011.09.026.

[144] Harii K, Asato H, Yoshimura K, Sugawara Y, Nakatsuka T, Ueda K. One-stage transfer of the latissimus dorsi muscle for reanimation of a paralyzed face: a new alternative. *Plast Reconstr Surg*. 1998;102(4):941-951. doi:10.1097/00006534-199809040-00001.

[145] Takushima A, Harii K, Asato H, Kurita M, Shiraishi T. Fifteen-year survey of one-stage latissimus dorsi muscle transfer for treatment of longstanding facial paralysis. *J Plast Reconstr Aesthetic Surg JPRAS*. 2013;66(1):29-36. doi:10.1016/j.bjps.2012.08.004.

[146] Bianchi B, Copelli C, Ferrari S, Ferri A, Bailleul C, Sesenna E. Facial animation with free-muscle transfer innervated by the masseter motor nerve in unilateral facial paralysis. *J Oral Maxillofac Surg Off J Am Assoc Oral Maxillofac Surg*. 2010;68(7):1524-1529. doi:10.1016/j.joms.2009.09.024.

[147] Guelinckx PJ, Sinsel NK. Muscle transplantation for reconstruction of a smile after facial paralysis past, present, and future. *Microsurgery*. 1996;17(7):391-401. doi:10.1002/(SICI)1098-2752 (1996)17:7<391::AID-MICR9>3.0.CO;2-J.

[148] Bianchi B, Copelli C, Ferrari S, Ferri A, Sesenna E. Successful salvage surgery after treatment failures with cross graft and free muscle transplant in facial reanimation. *J Cranio-Maxillo-fac Surg Off Publ Eur Assoc Cranio-Maxillo-fac Surg*. 2012;40(2):185-189. doi:10.1016/j.jcms.2011.03.016.

[149] Horta R, Silva P, Silva A, et al. Facial reanimation with gracilis muscle transplantation and obturator nerve coaptation to the motor nerve of masseter muscle as a salvage procedure in an unreliable

cross-face nerve graft. *Microsurgery.* 2011;31(2):164-166. doi:10. 1002/micr.20844.

[150] Chuang DC-C. Free tissue transfer for the treatment of facial paralysis. *Facial Plast Surg FPS.* 2008;24(2):194-203. doi:10.1055/ s-2008-1075834.

[151] Harrison DH, Grobbelaar AO. Pectoralis minor muscle transfer for unilateral facial palsy reanimation: an experience of 35 years and 637 cases. *J Plast Reconstr Aesthetic Surg JPRAS.* 2012;65(7):845-850. doi:10.1016/j.bjps.2012.01.024.

[152] Liu A-T, Lin Q, Jiang H, et al. Facial reanimation by one-stage microneurovascular free abductor hallucis muscle transplantation: personal experience and long-term outcomes. *Plast Reconstr Surg.* 2012;130(2):325-335. doi:10.1097/PRS.0b013e3182589d27.

[153] Alagöz MS, Alagöz AN, Comert A. Neuroanatomy of Extensor Digitorum Brevis Muscle for Reanimation of Facial Paralysis. *J Craniofac Surg.* Published online 2011. doi:10.1097/SCS. 0b013e318232a806.

[154] Tan ST, Staiano JJ, Itinteang T, McIntyre BC, MacKinnon CA, Glasson DW. Gold weight implantation and lateral tarsorrhaphy for upper eyelid paralysis. *J Cranio-Maxillo-fac Surg Off Publ Eur Assoc Cranio-Maxillo-fac Surg.* 2013;41(3):e49-53. doi:10.1016/j. jcms.2012.07.015.

[155] Liebau J, Schulz A, Arens A, Tilkorn H, Schwipper V. Management of lower lid ectropion. *Dermatol Surg Off Publ Am Soc Dermatol Surg Al.* 2006;32(8):1050-1056; discussion 1056-1057. doi:10.1111/j.1524-4725.2006.32229.x.

[156] Terzis JK, Karypidis D. Blink restoration in adult facial paralysis. *Plast Reconstr Surg.* 2010;126(1):126-139. doi:10.1097/PRS. 0b013e3181dbbf34.

[157] Boahene KDO. Dynamic muscle transfer in facial reanimation. *Facial Plast Surg FPS.* 2008;24(2):204-210. doi:10.1055/s-2008-1075835.

[158] Gs K, G M. Endoscopic forehead and brow lift. *Facial Plast Surg FPS.* 2009;25(4):222-233. doi:10.1055/s-0029-1242034.

[159] Verpaele A, Tonnard P, Gaia S, Guerao FP, Pirayesh A. The third suture in MACS-lifting: making midface-lifting simple and safe. *J Plast Reconstr Aesthetic Surg JPRAS*. 2007;60(12):1287-1295. doi:10.1016/j.bjps.2006.12.012.

[160] Alex JC, Nguyen DB. Multivectored suture suspension: a minimally invasive technique for reanimation of the paralyzed face. *Arch Facial Plast Surg*. 2004;6(3):197-201. doi:10.1001/archfaci.6.3.197.

[161] *Rehabilitation of Long-standing Facial Nerve Paralysis With Percutaneous Suture–Based Slings*. ResearchGate. doi:10.1001/archfaci.9.3.205.

[162] Capone RB, Sykes JM. The Effect of Rhytidectomy on the Nasal Valve. *Arch Facial Plast Surg*. 2005;7(1):45-50. doi:10.1001/archfaci.7.1.45.

[163] Rosson GD, Redett RJ. Facial palsy: anatomy, etiology, grading, and surgical treatment. *J Reconstr Microsurg*. 2008;24(6):379-389. doi:10.1055/s-0028-1082897.

[164] Alexander AJ, de Almeida JR, Shrime MG, Goldstein DP, Gilbert RW. Novel outpatient approach to lower lip reanimation using a palmaris longus tendon sling. *J Otolaryngol - Head Neck Surg J Oto-Rhino-Laryngol Chir Cervico-Faciale*. 2011;40(6):481-488.

[165] Husseman J, Mehta RP. Management of synkinesis. *Facial Plast Surg FPS*. 2008;24(2):242-249. doi:10.1055/s-2008-1075840.

[166] Salles AG, Toledo PN, Ferreira MC. Botulinum toxin injection in long-standing facial paralysis patients: improvement of facial symmetry observed up to 6 months. *Aesthetic Plast Surg*. 2009;33(4):582-590. doi:10.1007/s00266-009-9337-9.

[167] Pourmomeny AA, Zadmehre H, Mirshamsi M, Mahmodi Z. Prevention of synkinesis by biofeedback therapy: a randomized clinical trial. *Otol Neurotol Off Publ Am Otol Soc Am Neurotol Soc Eur Acad Otol Neurotol*. 2014;35(4):739-742. doi:10.1097/MAO.0000000000000217.

[168] Diels HJ. Facial paralysis: is there a role for a therapist? *Facial Plast Surg FPS*. 2000;16(4):361-364. doi:10.1055/s-2000-15546.

[169] Diels HJ. Treatment of facial paralysis using electromyographic feedback--a case study. *Eur Arch Oto-Rhino-Laryngol Off J Eur Fed Oto-Rhino-Laryngol Soc EUFOS Affil Ger Soc Oto-Rhino-Laryngol - Head Neck Surg*. Published online December 1994:S129-132. doi:10.1007/978-3-642-85090-5_39.

[170] Lindsay RW, Robinson M, Hadlock TA. Comprehensive facial rehabilitation improves function in people with facial paralysis: a 5-year experience at the Massachusetts Eye and Ear Infirmary. *Phys Ther*. 2010;90(3):391-397. doi:10.2522/ptj.20090176.

[171] Beurskens CHG, Heymans PG. Positive effects of mime therapy on sequelae of facial paralysis: stiffness, lip mobility, and social and physical aspects of facial disability. *Otol Neurotol Off Publ Am Otol Soc Am Neurotol Soc Eur Acad Otol Neurotol*. 2003;24(4):677-681. doi:10.1097/00129492-200307000-00024.

[172] Beurskens CHG, Heymans PG. Mime therapy improves facial symmetry in people with long-term facial nerve paresis: a randomised controlled trial. *Aust J Physiother*. 2006;52(3):177-183. doi:10.1016/s0004-9514(06)70026-5.

[173] Beurskens CHG, Heymans PG, Oostendorp RAB. Stability of benefits of mime therapy in sequelae of facial nerve paresis during a 1-year period. *Otol Neurotol Off Publ Am Otol Soc Am Neurotol Soc Eur Acad Otol Neurotol*. 2006;27(7):1037-1042. doi:10.1097/01.mao.0000217350.09796.07.

In: Cranial Nerves
Editor: Thomas M. Yi

ISBN: 978-1-53618-823-3
© 2021 Nova Science Publishers, Inc.

Chapter 4

VAGUS NERVE: THE CLINICAL IMPORTANCE IN THE METABOLIC DISORDERS

Berrin Zuhal Altunkaynak*, PhD,
Işınsu Alkan, PhD and Cengiz Baycu, PhD
Department of the Histology and Embryology,
Medical Faculty of İstanbul Okan University, İstanbul, Turkey

ABSTRACT

The vagus nerve is the 10th cranial nerve with motor functions responsible for the innervations of the outer ear canal, pharynx, larynx, heart, lung, gastrointestinal tract, stomach, pancreas and liver. The vagus nerve can be evaluated as a regulator of body metabolism by receiving signals from the brain. Signals from the hypothalamus, the master chief of the body, are transmitted by the vagus to most of the peripheral organs.

The bridge function between the brain and peripheral organs causes the vagus to be used as a treatment tool in metabolic disorders. The treatments are based on the vagus nerve as stimulation and blockade can

---

* Corresponding Author's E-mail: berrinzuhal@gmail.com.

be used in obesity, neural and metabolic diseases and diabetes. While the vagal stimulation is used in diseases such as obesity and blood sugar regulation; vagal blockade is used in the treatment of obesity, metabolic and neuronal diseases. The speed of the effect of vagal stimulation on each organ is very slow, especially in cases where biochemical reactions occur. For fast action, vagal blockage can be preferred. In addition, vagal effects on the pancreas in the gastrointestinal tract could initiate the inflammatory pathways. Although it does not receive direct innervation by the vagus; these pathways can induce same changes in the spleen. When planning vagal treatments, it is necessary to consider whether the treated organ which is innervated with the left or right branch of the vagus nerve. Another important point of the planning is which regions of these organs are innervated by which branch.

For example, a planned study to evaluate the effects of the left cervical vagus on the lung will not yield any results. Therefore, vagus nerve should be stimulated from the correct segment and branch in order to perform an effective treatment. In this section, the effects of vagal stimulation and blockade on different organs and their effects on metabolic diseases will be discussed.

Keywords: vagal stimulation, vagal blockade, metabolic diseases, obesity, vagus nerve

1. INTRODUCTION

There are 12 nerves in the human body that come out of the brain. These nerves are called cranial nerves. The 12 cranial nerves in the body exit the brain in pairs, helping to connect the brain with other parts of the body such as the head, neck, chest and abdomen. The cranial nerves are classified using Roman numerals according to their location. Some of them send sensory information to the brain, including details about smells, sights, tastes, and sounds. These nerves are known as sensory nerves. On the other hand, another group of the cranial nerves control the movement of various muscles and the function of certain glands. These are known as motor nerves. Some cranial nerves have sensory or motor functions, while others have both. The vagus nerve is located in the autonomic nervous system as the longest and most important nerve (Snell R, 2010).

In the modern life, chronic stress, traumas and injuries increase the activity of the sympathetic nervous system and disrupt the balance of the body, the vagus nerve is a part of the autonomic nervous system and the most important antagonist of the sympathetic nervous system to relax, reach mental calmness, fight stress and provides to feel happy for a better quality of life (Guyton and Hall, 2013). Many people are unaware how the vagus nerve is important in their body. In this section, it is aimed to give information about the general anatomical and physiological features of the vagus nerve, as well as its clinical effects on the metabolism.

2. ANATOMY OF THE VAGUS NERVE

The vagus nerve, the most important member of the parasympathetic system and the longest of the cranial nerves, extends to very large areas. Like other pairs of heads, it continues not only to the head and face area, but also to the Thorax and the abdomen, and by forming plexuses around the organs there, the vagus nerve acts as a bridge that provides the relationship between the cranial and sacral autonomic centers and it is the strongest parasympathetic nerve of the vegetative system. With this condition, it is a strong antagonist of the sympathetic system. It contains different fibers according to its spreading area and its function in these areas, these upper parts of the esophagus are motor fibers for the striated muscles of the pharynx, larynx and soft palate, the lower parts of the esophagus are parasympathetic fibers for the smooth muscles of the stomach and intestines and the bronchi. The fibers that go up to the heart and have a slow slowing effect are the sensory fibers that extend from the various organs of the abdomen and chest to the posterior outer part of the eardrum, especially around the vessels near the heart (Paulsen and Waschke, 2015). The branches of vagus nerve in the body are examined according to their location and innervation numbers. According to this, it is divided into four main sections as head, neck, chest and abdomen. All branches of the vagus nerve were illustrated in Figure 1.

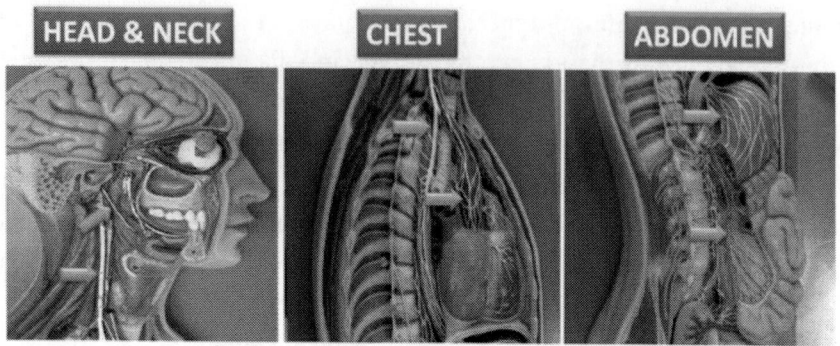

Figure 1. The branches of the vagus nerve (arrows) and the structures that innervated by the vagus nerve in head & neck, chest and abdomen regions.

The head part gives branches in the dura mater and is particularly sensitive to pain and heat, dissolving in the skin of the posterior face of the auricle and the external auditory canal. The neck portion extends downward surrounded by the common connective tissue sheath with the arteria carotis interna and arteria carotis communis and vena jugularis and is inserted with them into the apertura thoracis superior. The abdominal section contains sensory and parasympathetic fibers. In this part, the fibers that make up the vegetative part of the vagus and they are more dominant because the senses received from the abdominal organs are generally transmitted through the sympathetic pathways. They transmit to the nervous system (Ortug G, 2015).

The abdominal part of the vagus begins after the diaphragm passes, and this part consists of the oesophagus and the bronchioles that move towards the stomach. The two main branches, the truncus vagalis, which are separated at the oesophagus level and continue to the anterior and posterior abdominal parts, and together with the vagus fibers coming to the abdominal cavity, the nervus splanchnicus and truncus sympathetic fibers form an important neural network. Branches departing from this plexus extend to different abdominal organs and make new plexuses around these organs. Thus, parasympathetic and sympathetic innervations of the relevant organs are provided. Other anterior and posterior fibers extend to the anterior and posterior sides of the stomach where they branch into thin extensions and form a network on the front and back sides of the stomach and also the part of the large intestine up to the 1/3 transverse colon (Figure 1). It receives parasympathetic fibers from the vagus, the remaining part receives fibers via the sacral parasympathetic pathway (Berthoud & Neuhuber, 2000).

3. FUNCTIONS OF THE VAGUS NERVE

Vagus nerve; it is associated with the health of the whole body. It helps regulate many critical aspects of human physiology, including heart rate, blood pressure, sweating, digestion, and even speech. N. vagus is a mixed nerve with motor, sensory and parasympathetic functions. Motor function provides the innervation of the larynx (n. With recurrence), pharynx and oesophagus muscles. Its sensory fibers are responsible for the sensation of the larynx, pharynx, trachea, external auditory canal, and thoracic and abdominal vissers. Parasympathetic fibers, on the other hand, innervate thoracic and abdominal organs outside the sacral region. After exiting the skull, the root of the tongue gives branches to the abdominal organs such as the pharynx, larynx, trachea, oesophagus, heart and lungs, thoracic organs, stomach, liver and intestines. Efferent fibers show parasympathetic effect. It carries sensory information from the aortic arch. Additionally, it provides motor inputs for all laryngeal and pharyngeal muscles except the stylopharyngeal muscle and tensor palatini; provides sense of pharynx, ear and tympanic membrane (Agur and Dalley, 2009).

The vagus nerve, which is the most important member of the parasympathetic system, extends to very large areas, like other pairs of heads, it continues not only to the head and face area, but also to the thorax and the abdomen, and in a way, the vagus nerve branches off by forming plexses around the organs where it constitutes a relationship between the cranial and sacral autonomic centers. It acts as a bridge to the vegetative system as the strongest parasympathetic nerve and with this condition it is a strong antagonist of the sympathetic system. It contains different fibers according to its spreading area and its function in these areas, these upper parts of the esophagus are motor fibers for the striated muscles of the pharynx, the larynx and soft palate, the lower parts of the oesophagus, the parasympathetic fibers for the smooth muscles of the stomach and intestines and the bronchi, the secretory fibers for the glands of the digestive system and the respiratory tract. The sensory fibers of the vagus nerve having slowing effect are housed around the vessels, especially in the skin of the external auditory canal, the various organs of the abdomen and thorax, close to the heart, the outer face and the posterior part of the ear-drum (Figure 2; Moore et al., 2018).

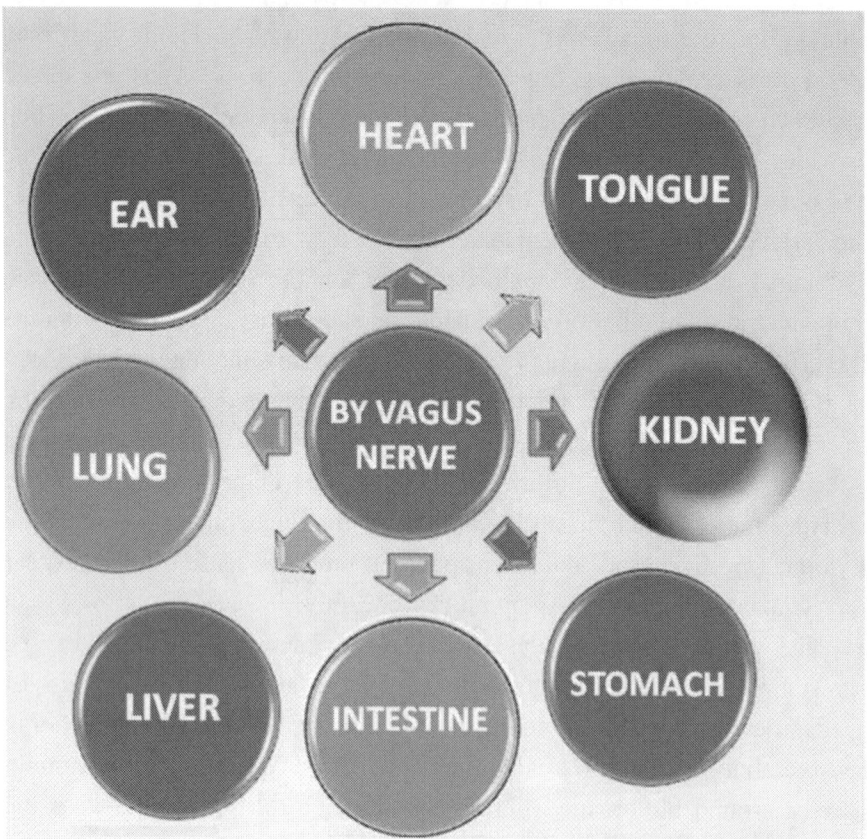

Figure 2. Functions of the vagus nerve in the body is illustrated. There are many functions of the vagus nerve. Hence, it innervates and controls important functions of the many organs.

4. CLINICAL INSIGHTS ABOUT VAGUS NERVE

The vagus nerve helps regulate all major body functions, from our breathing to our brain activities. It allows us to resist the difficulties of life and plays an important role in the regulation of gut movements, is also has great importance in the treatment of addictions, according to some studies. On the other hand, having a healthy vagus tone means a stronger immune system, positive emotions and psychological balance. Due to its close relationship with our health, the vagus nerve has become a frequently discussed topic in the medical world. Low vagal tone causes to inflammation, difficulties in the regulation of emotions and concentration, while many different symptoms such as depression, loneliness, anxiety, panic attack, gastroparesis (stomach emptying difficulty), tinnitus, weight gain and obesity, irritable bowel syndrome, abnormal heart rate, B12 deficiency. It is also closely related to health problems. In the lesion of the vagus nerve, hoarseness, difficulty in swallowing, and loss of the gag reflex occur. When a patient who opens his mouth is told to say "aaaa," we can evaluate the N. Vagus by the movement of the uvula; If the uvula slips to one side, there is a vagus lesion on the opposite side (Walker HK., 1990).

In order to evaluate the health of the vagus nerve, symmetry in the larynx and palate is checked with the help of a laryngoscope. In order to evaluate the parasympathetic function of N. vagus, oculo-cardiac reflex is checked. Bradycardia should occur after pressure on both eyes with fingers. Clinical symptoms may not be evident in unilateral lesions affecting the N. vagus, respiratory distress may occur only during exercise. There is respiratory distress and cyanosis due to paralysis of the laryngeal muscles. The voice is hoarse. Dysphagia and regurgitation are seen. Arrhythmia in the heart and impairment in gastrointestinal functions (mega esophagus, stomach dilatation) may result in death. The pharyngeal and oculo-cardiac reflex has disappeared (Walker HK., 1990).

Clinically, vagus paralysis is common. Occurs completely or partially; unilateral paralysis is affected by musculus cricoarytenoid and isolated paralysis is rare, this nerve paralysis can occur for different reasons. In these cases, the pharynx is usually affected and of course phonation and swallowing disorders are seen. Moreover, some problems occur in the esophagus, and brachial reflexes also disappear. A different picture is seen in the paralysis of the palate muscles. The main function of the palate muscles is to facilitate swallowing by closing the nasal cavity during swallowing. However, in the paralysis of these muscles, liquid foods are taken out of the nasal passage because muscle functions cannot be performed. In disorders involving the pharynx, shape changes occur in the palate calyces and their reflexes also disappear. In functional disorders of the vessels in the chest, tachycardia occurs and bradycardia occurs. In paralysis of the nerves close to the nasal muscles, breathing slows down. The clinical consequences of the paralysis of the branches close to the stomach are rarely mentioned. In such cases, it is stated that the feeling of hunger and thirst disappears. In addition, cramps of the muscles of the larynx are expressed as the clinical symptoms of conditions such as esophagitis, which are seen in the stimulation of the plexus pharyngeus and plexus esophagus. Unilateral disturbances in the vagus nerve are often seen as insignificant compared to total vagus disorders. The effects of serious strokes are always seen in both directions. The reason for this situation is in the nucleus area of the vagus, in the medulla oblongata or in the central vagus tracts. However, it should not be forgotten that such vagus paralysis may accompany neighbouring cranial nerves or other cerebral disorders (Simon and Mertens, 2009).

5. EFFECT OF VAGUS NERVE ON METABOLISM AND OBESITY

The vagus nerve is the nerve that communicates with the stomach and intestines and is an important part of the parasympathetic nervous system. This system physiologically works as the "rest and digest," this is the opposite of the sympathetic system that works with the "fight or flight" logic. Signals from the stomach and intestines to the brain via the vagus nerve create a sense of satiety. Signals from the brain through the vagus nerve affect digestion, the release of digestive enzymes and the motility of the digestive system. The most important connection between the vagus nerve and the digestive tract and the brain is the signals of hunger and satiety (Page et al., 2012). Foods that fill the stomach volume send satiety signals to the brain through the vagus nerve. This is the mechanism that allows the brain to perceive when satiety will be reached after eating. The purpose of the endoscopic gastric balloon applied in the temporary treatment of obesity is to create a mass effect in the stomach and to transmit satiety signals to the brain. In addition, the perception of nutrients in the intestines and neurotransmitters such as serotonin and ghrelin, which are also secreted from the intestines, send hunger and satiety signals to the brain via the vagus nerve (Figure 3, 5; Berthoud et al., 2011).

While some people are satiated with small amounts of food, those who are overweight often feel hungry even after a large meal. This situation is mostly related to the sensitivity of the vagus nerve. The "cafeteria diet" consisting of high fat-high carbohydrate foods may reduce the sensitivity of the vagus nerve, leading to obesity. The authors suggested that high-fat diet could potentially cause disruption of the intestinal barrier, such as sepsis and inflammatory bowel disease. Again, it is possible that very low fat and low-carbohydrate ketogenic diet and vagus nerve treatment may have an obesity-reducing effect.

Vagotomy diminishes obesity in cafeteria diet fed rats by decreasing cholinergic potentiation of insulin release (Balbo et al., 2016).

In the obesity condition, vagus nerve being insensitive to satiety signals. There are many studies that this is specifically due to diet. Successive diets lead to the vagus nerve to send satiety signals to the brain only when too much food is taken. If a poor diet affects the sensitivity of the vagus nerve, it can also cause inflammatory diseases such as irritable bowel disease or Crohn's. There are many data that depression and diabetes are also intestinal diseases. This explains why gut health is important in overall health. For this reason, stimulation of the vagus nerve, as can be predicted, leads to an earlier occurrence of satiety and weight loss both in experimental animal studies and in clinical trials on humans. Insulin release and glucose homeostasis in the liver, which are closely related to metabolism, are another mechanism regulated by the vagus nerve (Figure 3; Li et al., 2007).

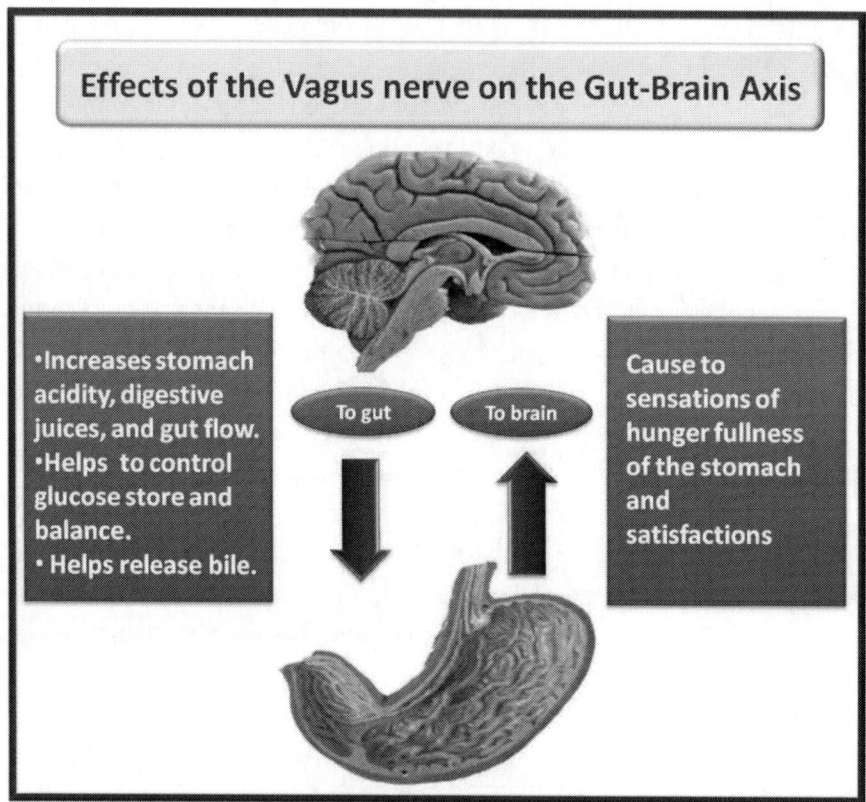

Figure 3. Effects of the vagus nerve on gut-brain axis are summarized. Vagus nerve transmit the signals from brain-to gut and from gut to brain.

6. Vagus Nerve as an Obesity Threating Agent

Since the vagus nerve has very important functions, medical science aims to focus on the effects of the vagus nerve on the body area to be treated. These medical treatments can be classified as using effective drugs on the parasympathetic nervous system, vagus nerve stimulation, and vagus nerve blockage. These types of treatment are mentioned below:

6.1. Drug Treatments

Nowadays, the effect of the parasympathetic nervous system on the human body is increased or suppressed by drugs (Katzung, 1998). For example, pyridostigmine, an acetylcholinesterase inhibitor, is used to restore heart tone in heart failure or after post-myocardial infarction. Statins, developed as drugs that lower blood lipid levels, partially regulate vagal activity and show pleiotropic effects (Bi et al., 2013). However, considering the effects of the vagus nerve on the digestive system, the use of drugs that regulate the activity of the vagus nerve alone is often not sufficient to treat metabolic diseases. In addition, the physiological functions of the vagus nerve on many cardio-respiratory systems have shown that a drug that affects the vagus nerve entirety may have many side effects. Therefore, many obesity drugs have been withdrawn due to side effects such as heart valve defects and pulmonary hypertension (Kang & Park, 2012). Therefore, a major challenge in developing a possible obesity drug through the vagus nerve will be to selectively target the organs that this drug will affect. Drugs for vagal afferents that affect only the abdominal organs may reduce the risk of cardio respiratory side effects. On the other hand, the development of drugs that act with tissue-organ specificity due to antigen-antibody communication may be effective in preventing unwanted side effects while developing a treatment. Such a drug should selectively bind to vagal afferent neurons that innervate the digestive tract, and may provide results that increase satiety, reduce the desire to eat, prevent the absorption of fats, or regulate thermoregulation by affecting the activity of only those cells. However, while this drug regulates the activities of the some nerve cells and it needs to be crossable from the blood-brain barrier and mustn't cause a negative effect on the central nervous system. Today, any drug with this feature has not been developed yet, but studies continue in this direction.

6.2. Neuromodulation Treatments by Vagus Nerve

Especially afferent pathways of the vagus carry out the activation of different stimulating or blocking pathways that will regulate weight gain and homeostasis. Vagal changes, especially in the gastrointestinal system, reduce food intake (de Lartigue, 2016). Although the mechanism of the effect on the intestines is not clearly explained, it is thought that food intake is regulated with the help of chemo and mechanoreceptors in the gastrointestinal system (Ritter, 2004; Berthoud, 2008a). The stimuli that are necessary for the regulation of the system and come from the liver, pancreatic beta cells, intestines, and stomach are transmitted to the brain by vagal afferents or hormonal pathways (Figure 5). Hormonal signals initiate the feedback mechanism in the hypothalamus and medulla region of the brain, while signals such as glucose, fat, protein from vagal afferents are responsible for stimulation in the cortex region outside the hypothalamus and medulla (Berthoud, 2008b). This shows that the association between the vagus and obesity creates different signals in each organ. On this aspect, researchers are in dilemma. There are two way of the treatment by the vagus nerve as stimulation and inhibition. But which of them is useful for obesity treatment is controversial (Figure 4).

6.2.1. Vagus Nerve Stimulation

Vagal nerve stimulation (VNS) was first mentioned by Soma Weiss in 1934 and has been used in the treatment of different diseases since then (Weiss and Ferris, 1934). The only approved stimulation technique VSS was first on a patient in 1988 has been tried. Afterwords, Penry and Dean reported a case and shared with the scientific world including successful results of the patient with epilepsy after a year from VNS in 1989 at the meeting of the Epilepsy Association (Penry and Dean 1990). Today, the most used Vagus Nerve Stimulator was developed in 1997 by Cyberonics® Inc. (Houston TX USA) and approved in the United States by American Food and Drug Administration (Food and Drug Administration; FDA). (Cyberonics® Inc, 2017). On July 2005, the stimulator has been accepted to be used in drug-resistant major depression by FDA. Other potential indications of the stimulator are chronic pain, migraine, nutrition disorders, obesity, multiple sclerosis and it can be considered as Alzheimer's disease (Handforth and Krahl, 2001; Yuan and Silberstein, 2015).

Nowadays, the effects of vagal stimulation, which is frequently preferred in the treatment of neurological diseases such as epilepsy and depression, on weight loss constitute the subject of many studies in the literature (Bodenlos et al., 2014; Bugajski et al., 2007; Roslin and Kurian, 2001; Zeiler et al., 2015). In studies, vagal nerve stimulation is performed with an apparatus or a stimulator that enables the left cervical vagus nerve to be affected (Bodenlos et al., 2014; Alkan et al., 2020). The apparatus stimulates the nerve at a period and time. The most easily stimulated fibers are afferent unmyelinated C-type fibers of the vagus nerve. In general, this stimulation is thought to inhibit food intake via the gut hormone cholecystokinin (CCK) and cause weight loss by activating vagal afferents that mediate satiety from the gastrointestinal tract (Schwartz, 2000). CCK is responsible for the contraction and release of the gallbladder, which is released after the intake and consumption of food (Roslin and Kurian, 2001). The bile released by this effect plays a role in fat digestion. Thus, release from the stomach to the duodenum is induced (Roslin and Kurian, 2001).

After stimulation, it was reported that weight loss increased and the feeling of hunger was suppressed, especially in obese animals (Bugajski et al., 2007; Alkan et al., 2020). In the previous study, vagal stimulation was applied to the subjects and it was observed that the food intake of the animals significantly decreased (Val-Laillet et al., 2010). In another study on male rats, the left cervical vagus of healthy rats was stimulated with the help of a device with a current of 1.5 mA and they showed that VNS especially affects adipose tissue without any difference in food intake. VNS increased energy production and caused a decrease in adipose tissue mass following the 2nd week of the application (Banni et al., 2012). In the same study, it is mentioned that VNS causes unesterified fatty acids to be mobilized from the adipose tissue without being stored in the liver, as well as reducing the amount of endocannnobinoids and thus reducing fat mass.

Appetite can be suppressed by reducing the activation of the receptors on the central nervous system with the decrease in the amount of endocannnobinoids. In another study on obese mini pigs, the effect of the vagal nerve stimulation on insulin and glucose values was examined, and the analysis was performed two weeks after the stimulator was placed. As a result, a relation between increased glucose metabolism and VNS was reported and they argued that chronic vagal stimulation significantly improved insulin sensitivity in diet-induced obesity by both peripheral and central mechanisms (Malbert et al., 2017).

In the another study, left cervical vagus of the obese rats was stimulated and the number of neurons in the animals' myenteric plexus, arcuate nucleus, ventromedial nucleus and dorsomedial nucleus were estimated (Alkan et al., 2020). At the end of the study, they showed that the weight of the animals in the stimulation group decreased by 8% compared to their weight at the beginning of the study (Alkan et al., 2020).

Although the effect of vagal stimulation on obesity is clear from animal studies, this effect is confused in clinical studies. Some studies argue that VNS causes to weight loss, while some studies suggest that it does not cause to any change (Burneo et al., 2002; Koren and Holmes, 2006). In the study by Pardo et al., vagal stimulation was applied for 2 weeks in the treatment of obese patients with drug resistant depression, and serious weight loss was observed in these patients despite there is not any diet or exercise program (Pardo et al., 2012). In one case report, an epileptic patient lost 12% weight after 15 months of VNS, but the weight remained constant even though VNS was given at a higher amplitude after a break (Khan et al., 2017). In the study in which 21 patients aged 35 years were analyzed, no relationship was found between weight loss and VNS (Koren and Holmes, 2006). A study by Johannessen found that the vagus nerve was gradually stimulated and food intake decreased by 10% in individuals (Johannessen et al., 2017).

The vagal stimulation mechanism is unclear. However, this effect is thought to be due to the change in energy metabolism and the regulation of glucose metabolism, creating a feeling of satiety. In a study by Sarr et al. on 503 patients, electrodes were placed on the vagus nerve and this stimulation changes the contractions of the stomach muscles and creates a feeling of fullness. In addition, in the same study, the researchers said that electrical stimulation changes intestinal secretions, reduces the exocrine secretion of the pancreas, and consequently, it may reduce food digestion and absorption (Sarr et al., 2012). In another study, it was reported that stimulated vagus nerve afferents reduce gastric contractions by 14% and increase gastric secretions by 10% (Wang et al., 1999).

Vagal stimulation has an effect on the brain by activating or inactivating pathways. The most important region in maintaining the balance of hunger and satiety in the brain is the hypothalamus nuclei where provide energy metabolism and appetite management by perceiving signals from peripheral organs (Alkan et al., 2017). Especially the paraventricular, arcuate, ventromedial, and dorsomedial nuclei located in the hypothalamus are frequently studied for food intake. In the study by Alkan et al., it was shown that vagus stimulation affects neuron density of the nuclei and weight loss, as well. The neuron density in the arcuate, ventromedial and dorsomedial nuclei were analyzed in the subjects who underwent vagal stimulation showed that the neuron density increased in the arcuate and ventromedial nuclei of the stimulated group. Otherwise the neuron density in the dorsomedial nucleus of the stimulated group was lower than in the control group (Alkan et al., 2020). The arcuate nucleus is very important in food intake and contains neurons responsible for neuropeptide Y secretion. It increases the release of NPY with the stimulation. The NPY secretion at the high levels may that is why of the increase in neuron density of this nucleus. There are many studies showing that lesions of the ventromedial nucleus are associated with eating and obesity (Gold, 1973; Storlien, 1985; Kiba et al., 1993). In the same study, an increased neuronal density was observed in the ventromedial nucleus. This increase seems the functions of the ventromedial nucleus and the feedback of stimulation and signals in the gastrointestinal tract such as peptide YY (Alkan et al., 2020).

6.2.2. Vagus Nerve Inhibition

Another treatment method performed through the vagus is vagotomy. Vagotomy is the cutting of one of its branches in order to eliminate the effect of VS (Dezfuli et al., 2018). Vagotomy is an essential component of the surgical treatment of duodenal and gastric peptic ulcer disease (PUD). Vagotomy has been used widely for many years to treat and prevent PUD. Surgery has begun to lose its indisputable place in ulcer treatment with the recognition of the role of anti-secretory drugs and Helicobacter pylori (H. pylori) in ulcer pathogenesis (Szabo et al., 2016). Studies on the vagus nerve are still ongoing. The effect of the vagus nerve especially on cardiac ,endothelial dysfunction and inflammations is remarkable (Chapleau et al., 2016).

In obesity, vagotomy is performed through sensory afferents. These vagotomy called sensory vagotomy, which is performed not through the removal of all fibers (Fox, 2012). There are studies suggesting that the sensory vagotomy applied has therapeutic effects on obesity (Ferrari et al., 2005; Melnyk and Himms-Hagen, 1995; Stearns et al., 2012). In a study by Ferrari et al., on obese male Zucker (fa / fa) mice, blocked upper intestinal digestion by vagotomy and changed food intake. In this study, there aren't significant difference was observed in insulin values. In another study, rats with all metabolic disorders including hyperphagic, obese, hyperinsulinemia, glucose intolerance, insulin resistance, hyperglycemia and hypertriglyceridemia showed a decrease in body weight and perigonadal fat stores twelve weeks after vagotomy. In the same study, it was reported that vagotomy changed glycemia, insulinemia, and insulin sensitivity without a cause in glucose tolerance (Balbo et al., 2016). This effect of vagotomy is thought to be achieved by the loss of cholinergic effect. Lack of cholinergic effect changes the insulin sensitivity of the pancreas. Thus, while insulinemia decreases, a decrease in fat storage is observed at the same time. In the study by Stearns et al., It was shown that sensory vagotomy did not cause a decrease in general body weight, but selectively caused a significant reduction in visceral fat in the abdominal region (Stearns et al., 2012). In the long process after vagotomy or damage to the vagus, vagus afferents regenerate themselves, and in some cases, they show adverse effects (Fox, 2012; Philip et al., 2000). In another study, subdiaphragmatic vagotomy showed that glucose tolerance improved in mice fed a low protein diet in the early stages of life; these results showed that vagotomy increases the plasma insulin concentration by decreasing insulin clearance and decreasing the expression of the insulin-degrading enzyme in the liver (Lubaczeusk et al., 2017).

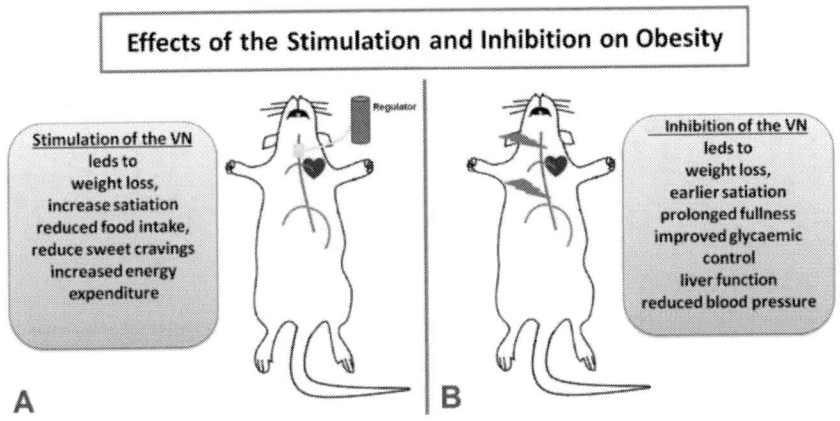

Figure 4. Effects of the stimulation and inhibition of the vagus nerve on obesity.

In another study, vagotomy was not performed, but 1 cm incision was made in the midline in the neck region, and left vagus nerves were crushed with a special clamp capable of applying 58 newtons pressure for 30 seconds. Weight was measured four weeks after the operation and a 14% decrease was found compared to per experiment (Alkan et al., 2020). When the neuron density in the arcuate, dorsomedial, and ventromedial nuclei of the hypothalamus were observed; neuron density was decreased in whole nuclei compared to the control group (Alkan et al., 2020). The arcuate and ventromedial nuclei generate neuronal signals that enable the activation of many pathways in food intake and energy metabolism. The decrease in neuronal density in these two nuclei can be explained by the blunting and degeneration of neurons that cannot receive stimulation due to damage. Neurons that do not get enough stimulation and secrete, degenerate. The dorsomedial nucleus is considered the center of satiety. In the same study, the decrease in neuronal density in the dorsomedial area suggests that inhibition proceeds through POMC, which is an important pathway for satiety. The immunohistochemical analyzes performed in the same study support this hypothesis. It was determined that the amount of

POMC increased and the amount of NPY decreased in the inhibition group (Alkan et al., 2020). All these results suggest that the inhibition process through the satiety center rather than the hunger centers. In the same study, when myenteric plexus volume in the inhibition group was examined, a volumetric decrease was observed (Figure 4).

7. Effect of the Vagal Treatments on the Organs

7.1. Stomach

It is an organ that contains vagal afferents intensely, and the myenteric plexuses between the stomach muscles are responsible for receiving the stimuli and transmitting them to the brain. Feedback mechanisms are initiated to regulate gastric food intake with the pressure created by the food intake the stomach or the stimulation of hormones such as ghrelin and leptin (Berthoud, 2008b; de Lartigue, 2016). Electrophysiological studies have shown that vagus afferents are stimulated depending on the volume of the food rather than the type (Mathis et al., 1998; de Lartigue, 2016).

7.2. Intestines

They play a role in the control of food intake, especially since they contain vagal endings in the mucosal area. Vagal endings do not extend into the lumen, but they associated with enteroendocrine cells, a type of cell located in the intestinal mucosa. As a result of this association, enteroendocrine cells that receive the signals of nutrients activate the hormonal pathways and stimulate vagal afferents (Mathis et al., 1998; de Lartigue, 2016). This stimulation may occur through cholecystokinin (Berthoud, 2008b; Patterson et al., 2002).

7.3. Liver

It has a very important role in glucose metabolism and the vagal afferents in the liver are stimulated by glucose as a result of fat and glucose oxidation (Berthoud, 2008b). The signals such as glucose level and energy storage create feedback on food intake or breakdown. The signals received provide activation in the liver not only through blocking food intake but also in terms of initiating metabolic pathways and increasing metabolic rate Berthoud, 2008b; de Lartigue, 2016). Studies on liver afferents have shown that changes in these nerve fibers, increase the metabolic rate and cause weight loss by preventing fat accumulation. (Berthoud, 2008b; Uno et al., 2006; Langhans, 2003; Gao et al., 2015). In a study in which the hepatic vagus branch was removed by vagotomy, they showed that while there was no change in glycogen and ATP levels, even if food intake did not change, weight loss and energy expenditure increased, decreased fat storage (López-Soldado et al., 2017). All these studies show that inhibition of the hepatic vagus is important in preventing obesity.

7.4. Pancreas

This organ, which is responsible for the regulation of glucagon and insulin balance, plays an active role in obesity. In the literature, there is a study that is mentioned that the pancreas is stimulated by vagal afferents coming from the liver, stomach, and intestines (Berthoud, 2008b). Stimulation of these vagal afferents affects pancreatic beta cells (Rohner-Jeanrenaud and Jeanrenaud, 1985; Rohner-Jeanrenaud et al., 1983). Vagal nerve stimulation activates multiple pathways in the beta cells of the pancreas, which release various neurotransmitters, and regulates glucose homeostasis (Yamomato et al., 2017). Stimulation or blockade of the vagal nerves initiates different pathways and changes insulin and glucagon secretion.

In therapy or treatment; there is a critical point that the researcher should not forget. Besides the positive effects of electrical stimulation of the nerve, it has some side effects. In addition to the disruption of the existing mechanism, electrical stimulation can cause dulling of physiological responses or other potential reactions (Yao et al., 2018). It is also clear that electrical stimulation can lead to damage to surrounding tissues.

Summarily, the molecular signals affect the gastro-intestinal system organs and interactions the organs and brain control the metabolic pathways. The signals transmitting by the vagus nerve or blood vessels were shown in Figure 5.

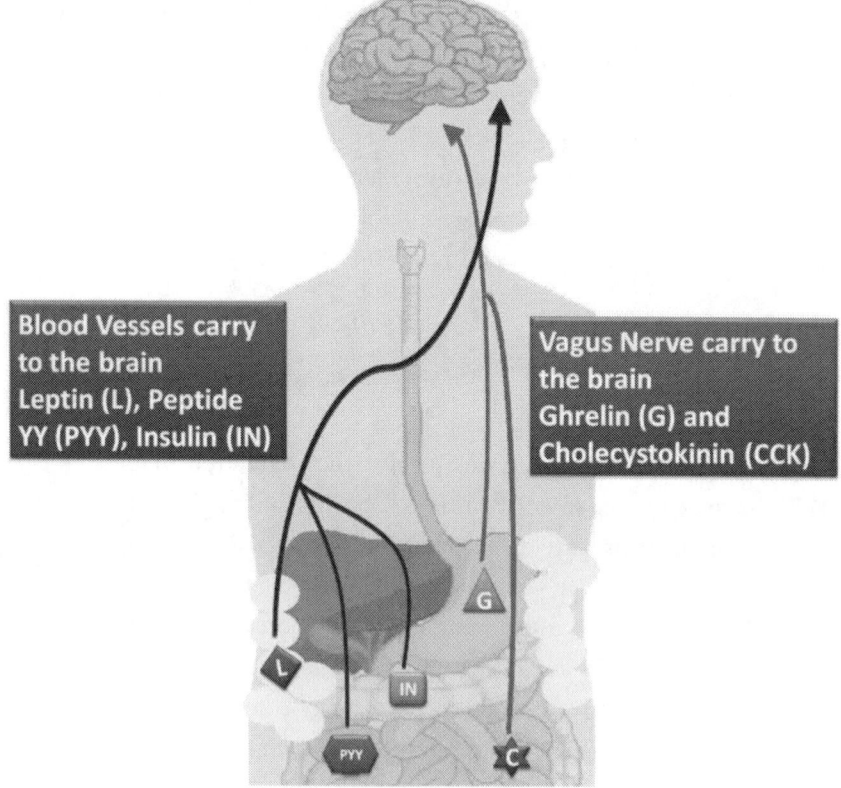

Figure 5. The molecular signals affecting metabolic pathways. The signals are transmitted by the vagus nerve or blood vessels.

8. EFFECT OF THE VAGAL TREATMENS ON THE MOLECULAR SIGNALS

8.1. Vagus nerve, BDNF, and Obesity

Brain-derived neurotrophic factor (BDNF) is responsible for the neuronal survival, synaptic healing and synaptic correlation localized on chromosome 11. It is a neurotrophin member of the neuron growth factor (NGF) family that plays a role in food intake, insulin sensitivity, and regulation of energy metabolism (Hayman et al., 1991; Angelucci et al., 2005; Brunoni et al., 2008). BDNF is synthesized in neurons, immune cells, adipocytes, endothelial cells (Pandit et al., 2020). The main role of BDNF on appetite is to create a feeling of satiety, and there are many studies showing the blockade of the BDNF gene cause to obesity (Figure 6; An et al., 2015; Banni et al., 2012). BDNF treatment shown to increase insulin sensitivity as it reduces glucose and insulin levels in diabetic mice, as well as reducing food intake, which causes loss of body weight (Wang et al., 2017). A study on the obese mice examined BDNF levels and found that BNDF levels were decreased in the obese compared to the control group (Jin et al., 2015). In a study on homozygous BDNF knockout mice, it was shown that animals with this mutation had deficiencies in both satiety and satiety, which led to an increase in meal size and frequency and a decrease in vagal signalling from the intestine to the brain (Fox et al., 2012). On the other hand, another study states that decreased BDNF count increases vagal innervation on the intestines and creates a feeling of satiety (Biddinger and Fox, 2014).

Vagus nerve stimulation is frequently used in diseases such as depression and epilepsy; furthermore, decreased BNDF levels are effective in the development of depression (Schlaepfer et al., 2008). Studies in this context have shown that VNS causes increased BDNF expression in the hippocampus (Follesa et al., 2007; Furmaga et al., 2012). Biggio et al. showed that VNS applied for the treatment of depression increased the BDNF expression in the hippocampus by the immunostaining method

(Biggio et al., 2009). As a result of this study, it was observed that the number of BDNF positive cells increased in the hippocampus CA3 region.

In a study, the obesity model was created on mice were exposed to NVS for 4 weeks and showed that BNDF expression increased in the ventromedial and paraventricular nuclei. This situation caused BDNF to accelerate energy metabolism and to weight loss in subjects (Banni et al., 2012). Although the hypothalamic ventromedial nucleus and dorsomedial nucleus express a small amount of BDNF playing a key role in pathways such as food intake and energy metabolism (Schwartz and Mobbs, 2012). Increasing the BDNF level in the brain by vagal stimulation and BDNF expression with intestinal stimulation suggests that it can be used for treatment in obesity.

8.2. Vagus Nerve, NPY and Obesity

Neuropeptide Y (NPY) is a protein secreted from neurons located in the arcuate nucleus of the hypothalamus. It is involved in the regulation of many neuroendocrine activities, especially appetite, autonomic functions, learning, stress response, sexual and motor behaviours (Alkan et al., 2017). After synthesized in the arcuate nucleus, NPY is transported to the paraventricular nucleus and binds to the receptors located here. Intracellular signals initiated by this binding regulate nutrional uptake stimulating by leptin, ghrelin and adipose signals (Figure 6; Alkan et al., 2017). The role of NPY in food intake had shown in many studies (Alkan et al., 2017; Bannon et al., 2000; Woods et al., 1998). The expression of NPY in the neurons of the arcuate nucleus caused to release of GABA vesicles and initiate appetite pathways (Sohn et al., 2015). It is also known that dorsomedial NPY signals descending to the dorsal motor nucleus of the vagus modulate hepatic insulin sensitivity to control hepatic glucose production in rats (Li et al., 2016). In a study in which NPY was inhibited, it was shown that descending signals were silenced, leading to an increase in hepatic vagal innervation (Li et al., 2016). This has altered the blood glucose level. In many studies conducted in mice, the association between the increase in NPY level and the increase in fat in the abdominal area was shown and it was stated that the selective effect of NPY on carbohydrate was more than its effect on fat (Gehlert, 1999). In the study by Alkan et al., NPY expression in the brain of obese rats was investigated by immunohistochemical analysis and it was observed that the number of NPY positive cells was higher in obese rats than in the control. Considering the appetite-enhancing function of NPY/AgRP neurons of arcuate nucleus in the hypothalamus supports the hypothesis. The NPY/AgRP neuron pathway proceeds through the NPY and the melanocortin receptor. In addition, AgRP increases potassium flow by affecting ion channels and produces hyperpolarization in the paraventricular nucleus (Ueno and Nakazato, 2016).

In the literature, there are few studies examining the role of NPY in vagus stimulation and inhibition in the treatment of obesity. However, when it is considered that the hypothalamic areas in the brain are affected in approaches that affect the vagus, it will also cause changes in NPY levels. In the literature, and in 2020. In the study performed by Alkan et al., vagus stimulation was applied in a group of rats, whose obesity model was formed, and vagus inhibition was applied in another group and NPY levels in the hypothalamic areas were examined immunohistochemically (Alkan et al., 2020). As a result of the analyzes, it was observed that the number of NPY positive cells decreased in both groups, while this number was quite low in the inhibition group (Alkan et al., 2020). In the same study, the neuron density in the arcuate nucleus was examined and it was observed that the density in the inhibition group decreased compared to stimulation (Alkan et al., 2020). In this case, it is quite possible that NPY is observed less in the inhibition group. In another study in the literature, hepatic vagotomy eliminated the inhibitory effect of dorsomedial NPY knockdown on hepatic glucose production, but they showed that this glycemic effect was not affected by vagal differentiations (Li et al., 2016).

The lack of studies in the literature reveals the need for further analysis of NPY in order to understand the pathway through which both stimulation and inhibition proceed. It is thought that explaining this pathway will constitute a important data in terms of obesity treatment.

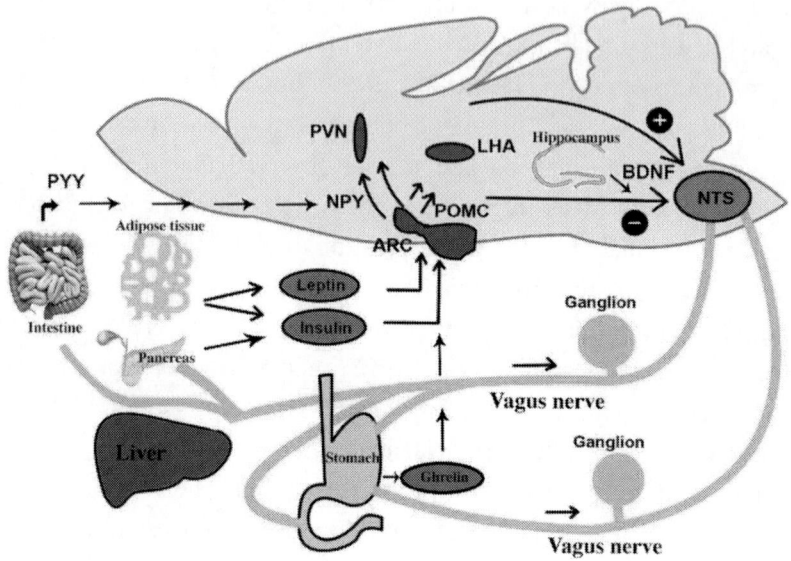

Figure 6. Food Intake Control Mechanisms. Vagus afferents stimulate peripheral organs such as liver, stomach, pancreas, intestines, and they regulate nutrient intake by activating different pathways in the hyphalamus nuclei and hippocampus in the brain. Stimulation of the ghrelin secreted from the stomach and signals of the myenteric plexus causes activation of the POMC neurons in the arcuate nucleus . The synthesized POMC passes to the lateral hypothalamic area and causes a feeling of satiety. Likewise, BNDF expression increases in the hippocampus area. Peptide YY secreted from the intestines stimulated by vagus afferents initiates the POMC pathway in the arcuate nucleus. On the other hand, the leptin and insulin synthesized from adipose tissue and pancreas, cause activation of NPY neurons in the arcuate nuclues and the synthesized NPY transfers to the paraventvricular nucleus, causing a feeling of appetite and hunger. (NPY: Neuropeptide Y, PVN: paraventricular nucleus, ARC: Arcuate nucleus, LHA: Lateral hypothalamic area, NTS: Neuronal transfer system, BDNF: Brain-derived neurotrophic factor, PYY: Peripheral peptide YY).

8.3. Vagus Nerve, POMC and Obesity

Proopiomelanocortin (POMC) is a protein that has a role in appetite suppression released from neurons localized in the arcuate nucleus in the hypothalamus (Milington, 2007; Sohn, 2015). POMC neurons are responsible for the production of different tissue-specific peptides through a series of enzymatic steps that produce melanocyte-stimulating hormones

(MSH), corticotropin (ACTH), and β-endorphin (Milington, 2007). It is thought that POMC's appetite suppressing effect proceeds through the release of α-MSH, which is an agonist at melanocortin-4 receptors (Sohn, 2015). In order for POMC to suppress appetite, expression must be induced by some signals (Figure 6). POMC neurons are located close to the capillary and are very sensitive to receiving hormonal signals. While the expression of POMC neurons by adipocyte-derived leptin is induced, it is suppressed by the stomach-derived ghrelin hormone (Cowley et al., 2011; Cowley et al., 2003). In addition, the melanocortinergic system, in which POMC is involved, also plays a role in the regulation of glucose metabolism (Santoro et al., 2004).

It is known that the secretions of POMC neurons are regulated by vagal afferents and efferents. Activation of POMC neurons in the arcuate nucleus was examined in rats with vagotomy; It has been shown that vagotomized animals significantly increase the number of POMC neurons with refeeding (Fekete et al., 2012). It is known that POMC neurons are activated by stimuli from vagus afferents, and it is quite possible that changes in the stimuli of the vagus will cause changes in the POMC pathway (Appleyard et al., 2005). The number of POMC positive cells in vagal stimulation and inhibition groups, which were mentioned in previous sections, were analyzed immunohistochemically. As a result of the study, the number of POMC positive cell in the vagal stimulation and inhibition group was higher than in the control group. The highest number of POMC positive cells was seen in the inhibition group (Alkan et al., 2020). In the same study, the decrease in the number of neurons in the dorsomedial nucleus in the inhibition group and the number of POMC positive cells increased in this group, while the decrease in the number of NPY positive shows that vagal inhibition proceeds through the center of satiety.

8.4. Vagus Nerve, Peptid YY and Obesity

Peripheral peptide YY (PYY) is an anorectic peptide that is responsible for nutrient uptake released from the distal gastrointestinal tract and has high affinity for the NPY Y2 receptor (Koda et al., 2005; Price and Bloom, 2014). They are secreted from L cells, a type of enteroendocrine cells in the gastrointestinal system. While it stimulates the ileal absorption, it slows down the absorption in the stomach and creates a feeling of satiety in the stomach (Wynne and Bloom, 2006; Strader and Woods, 2005; Hoentjen, 2001). PYY activates neurons in the arcuate nucleus and plays a role in regulating energy intake and food intake (Figure 6; Riediger et al., 2004). Plasma PYY levels begin to rise 15 minutes after taking to meal and reach a plateau phase in about 90 minutes. The satiety signal mediated by PYY inhibits NPY neurons and activates POMC neurons (Ueno et al., 2008). Peripheral PYY binds to Y2 receptors found in vagal afferents to transmit the satiety signal to the brain. PYY can also be found in selected neurons within the central nervous system, especially hypothalamus. Previous electrophysiological study has confirmed that PYY3–36 increases the firing rate of the gastric vagal afferent nerve (Koda et al., 2005).

In a study that examines food intake on rats with vagotomy; they applied peripheral PYY on rats and they showed that activation of arcuate nucleus neurons, food intake were interrupted (Abbott et al., 2005). In another study, exogenous administration of PYY decreased food intake in healthy rats but did not show this effect in rats with vagotomy (Koda et al., 2005). This result shows that PYY is carried by vagal afferents. PYY regulates food intake by affecting regions in the brain with the signals generated in the vagus. Serum peptide YY levels were examined after vagotomy in a study conducted on patients. They showed that serum PYY levels increased along with weight loss following vagotomy (Kim et al., 2012).

9. AFFECT OF THE VAGAL TREATMENTS ON THE DIABETES

Diabetes mellitus is a metabolic disorders characterized by a high blood glucose level (Ulubay et al., 2020). Irregular insulin secretion cause different metabolic problem. Diabetes mellitus can be seen in two forms, type 1, where insulin secretion is completely lost, and type 2, where beta cells are damaged and insulin secretion is impaired. The most common type is Type 2 (Keytsman et al., 2015). The important point in the treatment of diabetes is to reduce hyperglycemia and hyperlipidemia. For this reason, many drugs used have different toxic effects. This toxic effect has led to the development of new strategies. The innervation of the pancreas by the vagus nerve suggests that stimulation or blockade of vagus afferents may be effective in the treatment of diabetes mellitus. In addition, the fact that parasympathetic activation decreases hepatic glucose release and increases pancreatic insulin secretion in hyperglycemic conditions supports that vagal nerve stimulation may be effective in the treatment of type II diabetes (Meyers et al., 2016). In a study by Guo et al., they found decreased density in the myelinated nerve fibers in the vagus with morphological changes in diabetic patients (Guo et al., 1987). This situation reveals the association between the vagus and diabetes. There are studies in the literature investigating the effects of vagal stimulation on glucose and insulin metabolism (Malbert et al., 2017; Malbert, 2017). In a study by Malbert et al., stimulating electrodes were placed around the dorsal and ventral vagus and connected to a subcutaneous stimulus in 15 adult mini-pigs. Twelve weeks after surgery, glucose uptake and insulin sensitivity were measured using positron emission tomography. At the end of the measurements, they concluded that vagal stimulation was associated with increased glucose metabolism in the brain regions and significantly improved insulin sensitivity in diet-induced obesity (Malbert et al., 2017). In another study by Malbert, it was shown that insulin-mediated hepatic and skeletal muscle glucose uptake, which is disrupted by obesity, can be regulated by vagal stimulation (Malbert, 2017). Insulin sensitivity was restored with bilateral abdominal vagal stimulation. In the study by Yu et al., the association between glycosylated hemoglobin (HbA1c) level and vagus functional status was observed and its effect on blood glucose control in patients with type 2 diabetes mellitus was evaluated (Yu et al., 2020). As a result of the study, they observed that vagus function was impaired regardless of the HbA1c level. In a study on dogs, the effect of vagal stimulation on glucagon and insulin secretion was investigated, and an increase in glucagon and insulin levels was observed during vagal stimulation (Ahrén and Taborsky, 1986). Meyers et al. showed that vagal nerve stimulation in rats caused an increase in glucagon secretion in the pancreas (Meyers et al., 2016). Sobocki et al. examined the effects of vagal nerve stimulation on metabolism on pigs and showed that VNS caused lower IGF-I levels (Sobocki et al., 2006). In a study with vagal inhibition, vagal blockage was applied with the help of a special device in patients with diabetes mellitus, and it was shown that vagal inhibition in these patients caused significant weight loss, improvement in HbA1c levels and a decrease in blood pressure (Shikora et al., 2013).

CONCLUSION

The vagus, a cranial nerve, is responsible for initiating many organs and pathways in the body and acts as a bridge between the brain and peripheral organs in many metabolic diseases. For this reason, signal changes in the vagus nerve are very important in the treatment of metabolic diseases.

In this chapter, we present information about the general structure and anatomy of the vagus; was given some association between metabolic diseases especially obesity, diabetes mellitus, and vagal branches. All data in this chapter supported that the vagus nerve stimulation or inhibition has an important role for some signals for the regulation of secretion, activation, or inhibition different pathways.

REFERENCES

Abbott, C. R., M.,Monteiro, C. J., Small, A., Sajedi, K. L., Smith, J. R. C., Parkinson, M. A., Ghatei, S. R. Bloom. 2005. The inhibitory effects of peripheral administration of peptide YY(3-36) and glucagon-like peptide-1 on food intake are attenuated by ablation of the vagal-brainstem-hypothalamic pathway. Comparative Study. *Brain Res* 1044, no. 1:127-31. doi: 10.1016/j.brainres.2005.03.011.

Agur, A. M., A. F., Dalley. 2009. *The cranial nerves: Grants Atlas of Anatomy*. Philadelphia: Wilkins and Williams. pp. 814-815.

Ahrén, B., G. J. Jr., Taborsky. 1986. The mechanism of vagal nerve stimulation of glucagon and insulin secretion in the dog. *Endocrinology* 118, no. 4: 1551-1557.

Alkan, I., B. Z. Altunkaynak, E. G., Kivrak, A. A., Kaplan, G., Arslan. 2020. Is vagal stimulation or inhibition benefit on the regulation of the stomach brain axis in obesity? *Nutritional Neuroscience.* doi: 10.1080/1028415X.2020.1809875.

Alkan, I., B. Z., Altunkaynak, G., Altun, E., Erener 2017. The investigation of the effects of topiramate on the hypothalamic levels of fat mass/obesity-associated protein and neuropeptide Y in obese female rats. *Nutritional Neuroscience* 22: 1-10. Doi: 10.1080/1028415 X.2017.1374033.

An, J. J., G. Y., Liao, C. E., Kinney, N., Sahibzada, B., Xu. 2015. *Discrete BDNF Neurons in the paraventricular hypothalamus control feeding and energy expenditure.* 22: 175-188.

Angelucci, F., S., Brenè, A. A., Mathé. 2005. BDNF in schizophrenia, depression and corresponding animal models. *Mol Psychiatry* 10, no. 4: 345-352.

Appleyard, S. M., T. W., Bailey, M. W., Doyle, Y. H., Jin, J. L., Smart, M. J., Low, M. C., Andresen. 2005. Proopiomelanocortin neurons in nucleus tractus solitarius are activated by visceral afferents: Regulation by cholecystokinin and opioids. *Journal of Neuroscience* 25, no.14: 3578-3585.

Balbo, S. L., R. A., Ribeiro, M. C., Mendes, C., Lubaczeuski, A. C., Maller, E. M., Carneiro, M. L., Bonfleur, 2016. Vagotomy diminishes obesity in cafeteria rats by decreasing cholinergic potentiation of insulin release. *J Physiol Biochem* 72, 4: 625-633.

Balbo, S. L., R. A., Ribeiro, M. C., Mendes, C., Lubaczeuski, A. C. P. A., Maller, E. M., Carneiro, M. L., Bonfleur. 2016. Vagotomy diminishes obesity in cafeteria rats by decreasing cholinergic potentiation of insulin release. *J Physiol Biochem* 72, no. 4:625-633. doi: 10.1007/s13105-016-0501-9.

Banni, S., G., Carta, E., Murru, et al. 2012. Vagus nerve stimulation reduces body weight and fat mass in rats. *PLOS ONE* 7, no. 9: e44813.

Banni, S., G., Carta, E., Murru, et al. 2012. Vagus nerve stimulation reduces body weight and fat mass in rats. *PLOS ONE* 7, no: 9: e44813.

Bannon, A. W., J., Seda, M., Carmouche, J. M., Francis, M. H., Norman., B., Karbon, et al. 2000. Behavioral characterization of neuropeptide Y knockout mice. *Brain Res* 868: 79–87.

Berthoud, H. R., A. C., Shin, H., Zheng. 2011. Obesity surgery and gut-brain communication. *Physiol Behav* 105:106-119.

Berthoud, H. R. 2008a. Vagal and hormonal gut-brain communication: From satiation to satisfaction. *Neurogastroenterol Motil* 20, no. 1: 64–72.

Berthoud, H. R. 2008b. The vagus nerve, food intake and obesity. *Regul Pept* 149: 15–25.

Berthoud, H. R., W. L., Neuhuber. 2000. Functional and chemical anatomy of the afferent vagal system. *Auton Neurosci* 85:1–17.

Bi, X. Y., X., He, M., Zhao, X. J., Yu, W. J., Zang. 2013. Role of endothelial nitric oxide synthase and vagal activity in the endothelial protection of atorvastatin in ischemia/reperfusion injury. *J Cardiovasc Pharmacol* 61:391-400.

Biddinger, J. E., Fox, E. A. 2014. Reduced intestinal brain-derived neurotrophic factor increases vagal sensory innervation of the intestine and enhances satiation. *J Neurosci* 34, no. 31: 10379-10393.

Biggio, F., G., Gorini, C., Utzeri, P., Olla, F., Marrosu, I., Mocchetti, P., Follesa. 2009. Chronic vagus nerve stimulation induces neuronal plasticity in the rat hippocampus. *International Journal of Neuropsychopharmacology* 12, no. 9: 1209–1221.

Bodenlos, J. S., K. L., Schneider, J., Oleski, K., Gordon, A. J., Rothschild, S. L., Pagoto. 2014. Vagus nerve stimulation and food intake: Effect of body mass index. *Journal of Diabetes Science and Technology* 8, no. 3: 590–595.

Brunoni, A. R., M., Lopes, F., Fregni, 2008. A systematic review and meta-analysis of clinical studies on major depression and BDNF levels: implications for the role of neuroplasticity in depression. *Int J Neuropsychopharmacol* 11, no. 8: 1169-80. doi: 10.1017/S14611 45708009309.

Bugajski, A. J., K., Gil, A., Ziomber, D., Zurowski, W., Zaraska, P. J., Thor. 2007. Effect of long-term vagal stimulation on food intake and body weight during diet induced obesity in rats. *J Physiol Pharmacol.* 58: 5-12.

Burneo, J. G., E., Faught, R., Knowlton, R., Morawetz, R., Kuzniecky. 2002. Weight loss associated with vagus nerve stimulation. *Neurology* 59, no. 3: 463-464.

Chapleau M. W, D. L. Rotella, J. J. Reho, K. Rahmouni, H. M. Stauss. Chronic vagal nerve stimulation prevents high-salt diet-induced endothelial dysfunction and aortic stiffening in stroke-prone spontaneously hypertensive rats. *Am J Physiol Heart Circ Physiol* 2016; 311: H276-85.

Cowley, M. A., J. L., Smart, M., Rubinstei, M. G., Cerdán, S., Diano, T. L., Horvath, R. D., Cone, M. J., Low. 2001. Leptin activates anorexigenic POMC neurons through a neural network in the arcuate nucleus. *Nature* 411, no. 6836:480-484.

Cowley, M. A., R. G., Smith, S., M., Diano Tschöp, N. Pronchuk, K. L., Grove, C. J., Strasburger, M., Bidlingmaier, M., Esterman, M. L., Heiman, L. M., Garcia-Segura, E. A., Nillni, P., Mendez, M. J., Low, P., Sotonyi, J. M., Friedman, H., Liu, S., Pinto, W. F., Colmers, R. D., T. L., Cone Horvath. 2003. The distribution and mechanism of action of ghrelin in the CNS demonstrates a novel hypothalamic circuit regulating energy homeostasis. *Neuron* 37, no. 4:649-661.

Cyberonics® Inc: *VNS Therapy User Guide*. Houston: 2017.

Fekete, C., G., Zseli, P. S., Singru, A., Kadar, G., Wittmann, T., Fuzesi, R. M., Lechan, 2012. Activation of anorexigenic POMC Neurons during refeeding is independent of vagal and brainstem inputs. *Zenodo*. Accesed: 10.09.2020.

Ferrari, B., M., Arnold, R. D., Carr, et al. 2005. Subdiaphragmatic vagal deafferentation affects body weight gain and glucose metabolism in obese male Zucker (fa/fa) rats. *Am J Physiol Regul Integr Comp Physiol* 289: R1027–R1034.

Fetal, H. 2001. Effect of circulating peptide YY on gallbladder emptying in humans. *Scand J Gastroenterol* 36: 1086–1091.

Follesa, P., F., Biggio, G., Gorini, S., Caria, G., Talani, L., Dazzi, M., Puligheddu, F., Marrosu, G., Biggio. 2007. Vagus nerve stimulation increases norepinephrine concentration and the gene expression of BDNF and bFGF in the rat brain. *Brain Res* 1179: 28-34.

Fox, E. A. 2012. Treating diet-induced obesity: A new role for vagal afferents? *Digestive Diseases and Sciences* 57: 1115–1117.

Furmaga, H., F. R., Carreno, A., Frazer. 2012. Vagal nerve stimulation rapidly activates brain-derived neurotrophic factor receptor TrkB in rat brain. *PLoS One* 7, no. 5: e34844.

Gao, X., J. N., L., van der Veen, T., Zhu, M., Chaba, S., Ordoñez, D. P. Y., Lingrell, J. R. B., Koonen, Dyck, A., Gomez-Muñoz, D. E., Vance, R. L. Jacobs. 2015. Vagus nerve contributes to the development of steatohepatitis and obesity in phosphatidylethanolamine N-methyltransferase deficient mice. *J Hepatol* 62, no. 4: 913-920. doi: 10.1016/j.jhep.2014.11.026.

Gehlert, D. R. 1999. Role of hypothalamic neuropeptide Y in feedingand obesity. *Neuropeptides* 33: 329.

Gold, R. M. 1973. Hypothalamic obesity: The myth of the ventromedial nucleus. *Science*. 182: 488–490.

Guo, Y P, J. G., McLeod, J, Baverstock. 1987. Pathological changes in the vagus nerve in diabetes and chronic alcoholism. *J Neurol Neurosurg Psychiatry* 50, no. 11: 1449-1453. doi: 10.1136/jnnp.50.11.1449.

Guyton, A. C., J. E., Hall. 2013. *Textbook of medical physiology*. Çeviri editörü: Çavuşoğlu, H. Tıbbi Fizyoloji. 12. Edi. İstanbul: Nobel Tıp Kitabevleri. pp. 729-34.

Handforth, A., S. E., Krahl. 2001. Suppression of harmaline-induced tremor in rats by vagus nerve stimulation. *Movement Disorders* 16, 84-88.

Hyman, C., M., Hofer, Y. A., Barde, M., Juhasz, G. D., Yancopoulos, S. P., Squinto, R. M., Lindsay. 1991. BDNF is a neurotrophic factor for dopaminergic neurons of the substantia nigra. *Nature* 350 no. 6315: 230-232.

Jin, Y. J., P. J., Cao, W. H., Bian, M. E., Li, R., Zhou, L. Y., Zhang, M. Z., Yang. 2015. BDNF levels in adipose tissue and hypothalamus were reduced in mice with MSG-induced obesity. *Nutr Neurosci* 18, no. 8:376-382.

Johannessen, H., D., Revesz, Y., Kodama, N., Cassie, et al., 2017. Vagal blocking for obesity control: A possible mechanism-of-action. *Obes Surg* 27:177–185.

Katzung, B. G., 1998. *Basic and Clinical Pharmacology*, 7th ed. New York: Appleton and Lange.

Keytsman, C., P., Dendale, D., Hansen. 2015. Chronotropic incompetence during exercise in type 2 diabetes: aetiology, assessment methodology, prognostic impact and therapy. *Sports Med* 45, no. 7: 985-995.

Khan, F. A., M., Poongkunran, B., Buratto. 2017. Desensitization of stimulation-induced weight loss: A secondary finding in a patient with vagal nerve stimulator for drug-resistant epilepsy. *Epilepsy Behav Case Rep.* 8: 51-54.

Kim, H. H., M. I., Park, S. H., Lee, H. Y., Hwang, S. E., Kim, S. J., Park, W., Moon. 2012. Effects of vagus nerve preservation and vagotomy on peptide YY and body weight after subtotal gastrectomy. *World J Gastroenterol* 18, no. 30: 4044–4050.

Koda, S., Y., Date, N., Murakami, T., Shimbara, T., Hanada, K., Toshinai, A., Niijima, M., Furuya, N., Inomata, K., Osuye, M., Nakazato. 2005. The role of the vagal nerve in peripheral pyy3–36-induced feeding reduction in Rats. *Endocrinology* 146, no. 5: 2369–2375.

Koren, M. S., M. D., Holmes. 2006. Vagus nerve stimulation does not lead to significant changes in body weight in patients with epilepsy. *Epilepsy Behav.* 8, no. 1: 246-249.

Langhans, W., 2003. Role of the liver in the control of glucose-lipid utilization and body weight. *Curr Opin Clin Nutr Metab Care* 6: 449-455.

Li, C., E. S., Ford, L. C., Mc Guire, A. H., Mokdad, 2007. Increasing trends in waist circumference and abdominal obesity among US adults. *Obesity* 15: 216–224.

Li, l., C. b., La Serre, N., Zhang, L., Yang, H., Li, S., Bi. 2016. Knockdown of neuropeptide Y in the dorsomedial hypothalamus promotes hepatic insulin sensitivity in male rats. *Endocrinology* 157, no. 12: 4842-4852. doi: 10.1210/en.2016-1662.

López-Soldado, I., R., Fuentes-Romero, J., Duran, J. J., Guinovart. 2017. Effects of hepatic glycogen on food intake and glucose homeostasis are mediated by the vagus nerve in mice. *Diabetologia* 60: 1076–1083.

Lubaczeuski, C., L., Mateus Gonçalves, J. F., Vettorazzi, M. A., Kurauti, J. C., Santos-Silva, M. L., Bonfleur, A. C., Boschero, J. M. Costa-Júnior, E. M. Carneiro. 2017. Vagotomy reduces insulin clearance in obese mice programmed by low-protein diet in the adolescence. *Neural Plasticity* Article ID 9652978.

Malbert, C. H. 2017. Could vagus nerve stimulation have a role in the treatment of diabetes? *Bioelectronics in Medicine* 1, no. 1. doi.org/10.2217/bem-2017-0008.

Malbert, C. H., C., Picq, J. L., Divoux, C., Henry, M., Horowitz. 2017. Obesity-associated alterations in glucose metabolism are reversed by chronic bilateral stimulation of the abdominal vagus nerve. *Diabetes* 66, no. 4:848-857.

Mathis, C., Moran, T. H., Schwartz, G. J. 1998. Load-sensitive rat gastric vagal afferents encode volume but not gastric nutrients. *Am J Physiol Regul Integr Comp Physiol* 274: R280–R286.

Melnyk, A., J., Himms-Hagen. 1995. Resistance to aging-associated obesity in capsaicin-desensitized rats 1 year after treatment. *Obes Res* 3: 337–344.

Moore, K. L., A. F., Dalley, A. M., Agur. 2018. *Clinically Oriented Anatomy*, Lippincott Williams and Wilkin.

Ortug, G., 2015. *Kranial Sinirlerin Fonksiyonel Anatomisi*. Baysan Basım Evi. İstanbul. pp.84.

Page, A. J., E., Symonds, M., Peiris, L. A., Blackshaw, R. L., Young, R. L. 2012. Peripheral neural targets in obesity. *Br J Pharmacol* 166, 1537-1558.

Pandit, M., T., Behl, M., Sachdeva, S., Aror. 2020. Role of brain derived neurotropic factor in obesity. *Obesity Medicine* 17: 100189.

Pardo, J. V., S. A., Sheikh, M. A., Kuskowski, C., Surerus-Johnson, M. C., Hagen, J. T., Lee, B. R., Rittberg, D. E., Adson. 2007. Weight loss during chronic, cervical vagus nerve stimulation in depressed patients with obesity: An observation. *Int J Obes (London)* 31, no. 11: 1756-1759. doi: 10.1038/sj.ijo.0803666.

Patterson, L. M., H., Zheng, H. R., Berthoud. 2002. Vagal afferents innervating the gastrointestinal tract and CCKA-receptor immunoreactivity. *Anat Rec* 266: 10-20.

Paulsen, F., J., Waschke. 2015. Sobotta Atlas of Human Anatomy. Vol. 3: *Head, Neck and Neuroanatomy*. pp. 318.

Penry, J. K., J. C., Dean, 1990. Prevention of intractable partial seizures by intermittent vagal stimulation in humans: Preliminary results. *Epilepsia* 31, 2:40-43.

Phillips, R. J., E. A., Baronowsky, T. L., Powley. 2000. Regenerating vagal afferents reinnervate gastrointestinal tract smooth muscle of the rat. *J Comp Neurol* 421: 325–346.

Riediger, T., C., Bothe, C., Becskei, T. A. Lutz. 2004. Peptide YY directly inhibits ghrelin-activated neurons of the arcuate nucleus and reverses fasting-induced c-Fos expression: Comparative Study. *Neuroendocrinology* 79, no. 6:317-26. doi: 10.1159/000079842.

Ritter, R. C. 2004. Gastrointestinal mechanisms of satiation for food. *Physiol Behav* 81: 249 -273.

Rohner-Jeanrenaud, F., B., Jeanrenaud. 1985. A role for the vagus nerve in the etiology and maintenance of the hyperinsulinemia of genetically obese fa/fa rats *Int. J Obes* 9 no. 1: 71-75.

Rohner-Jeanrenaud, F., Hochstrasser, A. C., Jeanrenaud, B. 1983. Hyperinsulinemia of preobese and obese fa/fa rats is partly vagus nerve mediated. *Am J Physiol* 244, no. 4: E317-22. doi: 10.1152/ajpendo.1983.244.4.E317.

Roslin, M., M., Kurian. 2001. The use of electrical stimulation of the vagus nerve to treat morbid obesity. *Epilepsy Behav* 2:S11-S16.

Santoro, N., E. M., del Giudice, G., Cirillo, P., Raimondo, I., Corsi, A., Amato, A., Grandone, L., Perrone. 2004. An insertional polymorphism of the proopiomelanocortin gene is associated with fasting insulin levels in childhood obesity. *J Clin Endocrinol Metab* 89, no. 10: 4846-4849. doi: 10.1210/jc.2004-0333.

Sarr, M. G., C. J., Billington, R., Brancatisano, A., Brancatisano, J., Toouli, L., Kow, N. T., Nguyen, R., Blackstone, J. W., Maher, S., Shikora, D. N., Reeds, J. C., Eagon, B. M., Wolfe, R. W., O'Rourke, K., Fujioka, M., Takata, J. W., Swain, J. W., Morton, S., Ikramuddin, M, Schweitzer, B., Chand, R., Rosenthal, The EMPOWER Study Group. 2012. The EMPOWER study: Randomized, prospective, double-blind, multicenter trial of vagal blockade to induce weight loss in morbid obesity. *Obes Surg.* doi: 10.1007/s11695-012-0751-8.

Schlaepfer, T. E., C., Frick, A., Zobel, W., Maier, I., Heuser, M., Bajbouj, V., O'Keane, C., Corcoran, R., Adolfsson, M., Trimble, H., Rau, H. J., Hoff, F., Padberg, F., Müller-Siecheneder, K., Audenaert, D., Van den Abbeele, Z., Stanga, M., Hasdemir. 2008. Vagus nerve stimulation for depression: Efficacy and safety in a European study. *Psychol Med* 38, no. 5:651-61.

Schwartz, E., C. V., Mobbs. 2012. Hypothalamic BDNF and obesity: Found in translation. *Nature Medicine* 18: 496–497.

Schwartz, G. J. 2000. The role of gastrointestinal vagal afferents in the control of food intake: Current prospects. *Nutrition* 16: 866–873.

Shikora, S., J., Toouli, M. F., Herrera, B., Kulseng, H., Zulewski, R., Brancatisano, L., Kow, J. P., Pantoja, G., Johnsen, A., Brancatisano, K. S., Tweden, M. B., Knudson, C. J., Billington. 2013. Vagal blocking improves glycemic control and elevated blood pressure in obese subjects with type 2 diabetes mellitus. *Europe PMC*. 2013: 245683.

Simon, E., P., Mertens. 2009. Functional anatomy of the glossopharyngeal, vagus, accessory and hypoglossal cranial nerves. *Neurochirurgie* 55, no. 2: 132-135.

Snell, R., 2010. *Snell clinical neuroanatomy.* Lippincott Williams and Wilkins. pp. 352.

Sobocki, J., G., Fourtanier, J., Estany, P., Otal. 2006. Does vagal nerve stimulation affect body composition and metabolism? Experimental study of a new potential technique in bariatric surgery. *Europe PMC.* 139, no. 2: 209-216.

Sohn, J. W. 2015. Network of hypothalamic neurons that control appetite. *BMB Rep* 48, no. 4: 229–233.

Stearns, A. T., A., Balakrishnan, A., Radmanesh, S. W. Ashley, D. B., Rhoads, A., Tavakkolizadeh. 2012. Relative contributions of afferent vagal fibers to resistance to diet-induced obesity. *Dig Dis Sci* 57, no. 5: 1281–1290.

Storlien, L. H. 1985. The ventromedial hypothalamic area and the vagus are neural substrates for anticipatory insulin release. *Journal of the Autonomic Nervous System* 13, no. 4: 303-310.

Strader, A. D., S. C., Woods. 2005. Gastrointestinal hormones and food intake. *Gastroenterology* 128: 175–191.

Szabo IL, J Czimmer, G Mozsik. Cellular Energetical Actions of "Chemical" and "Surgical" Vagotomy in Gastrointestinal Mucosal Damage and Protection: Similarities, Differences and Significance for Brain-Gut Function. *Curr Neuropharmacol.* 2016;14(8):901-913.

Takayoshi K., T., Katsuaki, E., Osamu, S., Inoue. 1993. Ventromedial hypothalamic lesions increase gastrointestinal DNA synthesis through vagus nerve in rats. *Gastroenterology* 104: 475-448.

Ueno, H., H., Yamaguchi, M., Mizuta, M., Nakazato. 2008. The role of PYY in feeding regulation *Review Regul Pept* 145, no. 1-3:12-26. doi: 10.1016/j.regpep.2007.09.011.

Ueno, H., M., Nakazato. 2016. Mechanistic relationship between the vagal afferent pathway, central nervous system and peripheral organs in appetite regulation. *J Diabetes Investig* 7, no. 6: 812–818.

Ulubay, M, I., Alkan, K. K., Yurt, S., Kaplan. 2020. The protective effect of curcumin on the diabetic rat kidney: A stereological, electron microscopic and immunohistochemical study. *Acta Histochem* 122, no. 2 :151486. doi: 10.1016/j.acthis.2019.151486.

Uno, K., H., Katagiri, T., Yamada, et al. 2006. Neuronal pathway from the liver modulates energy expenditure and systemic insulin sensitivity. *Science* 312: 1656-1659.

Val-Laillet, D., A., Biraben, G., Randuineau, C. H., Malbert. 2010. Chronic vagus nerve stimulation decreased weight gain, food consumption and sweet craving in adult obese minipigs. *Appetite* 55, no. 2: 245-252.

Walker, H. K. 1990. Cranial nerves IX and X: The glossopharyngeal and vagus nerves In: Walker, H. K., W. D., Hall, J. W., Hurst (Ed), *Clinical methods, The history, physical, and laboratory examinations* 3rd edition Butterworths.

Wang, H., B., Wang, H., Yin, G., Zhang, L., Yu, X., Kong, H., Yuan, Xi., Fang, Q., Liu, Cu., Liu, L., Shi. 2017. Reduced neurotrophic factor level is the early event before the functional neuronal deficiency in high-fat diet induced obese mice. *Metab. Brain Dis* 32, no. 1: 247-257.

Wang, Z., H., Zheng, H. R., Berthoud. 1999. Funtional vagal input to chemically identified neurons in pancreatic ganglia as revealed by Fos expression. *Am J Physiol* 277: 958-964.

Woods, S. C., D. P., Figlewicz, L., Madden, D., Porte, A. J., R. J., Sipols Seeley. 1998. NPYand food intake: Discrepancies in the model. *Regul Pept.* 75–76:403–8.

Wynne, K., S. R., Bloom. 2006. The role of oxyntomodulin and peptide tyrosine–tyrosine (PYY) in appetite control. *Nature Clinical Practice Endocrinology & Metabolism* 2: 612–620.

Yamamoto, J., J., Imai, T., Izumi, Hi., Takahashi, Y., Kawana, K., Takahashi, S., Kodama, K., Kaneko, J., Gao, K., Uno, S., Sawada, T., Asano, V. V., Kalinichenko, E. A., Susaki, M., Kanzaki, H. R., Ueda, Y., Ishigaki, T., Yamada, H., Katagiri. 2017. Neuronal signals regulate obesity induced β-cell proliferation by FoxM1 dependent mechanism. *Nat Commun* 8: 1930. https://doi.org/10.1038/s41467-017-01869-7.

Yao, G., Kang, L., Li, Y., Long, H., Wei, C. A., Ferreira, J. J., Jeffery, Y., Lin, W., Cai. 2018. Effective weight control via an implanted self-powered vagus nerve stimulation device. *Nat Commun* 9: 5349.

Yu, Y., Hu, L., Xu, Y., Wu, S., Chen, Y., Zou, W., Zhang, M., Wang, Y., Gu, Y. 2020. Impact of blood glucose control on sympathetic and vagus nerve functional status in patients with type 2 diabetes mellitus. *Acta Diabetologica* 57: 141-150.

Yuan, H., S. D., Silberstein. 2015. Vagus nerve and vagus nerve stimulation, a comprehensive review: Part II. *Headache* 56, 259–266.

Zeiler, F. A., K. J., Zeiler, J., Teitelbaum, L. M., Gillman, M., West. 2015. VNS for refractory status epilepticus. *Epilepsy Res* 112:100-113.

INDEX

A

acetylcholinesterase, 127
acetylcholinesterase inhibitor, 127
acute intermittent porphyria, 92
adipose, 71, 78, 129, 138, 140, 149
adipose tissue, 129, 140, 149
adults, 60, 67, 68, 87, 150
amplitude, 76, 80, 82, 90, 130
anastomosis, 20, 86, 94, 112
anatomy, vii, viii, ix, 2, 3, 20, 25, 26, 29, 74, 82, 97, 115, 144, 146, 153
aneurysm, 25, 31
antibiotic, 88, 89
antiviral therapy, 102
apex, 47, 54, 55, 63, 64
aplasia, vii, 1, 14, 15
appetite, 131, 137, 138, 140, 153, 154
arteries, 9, 18, 19, 20, 21, 45, 60
arteriovenous malformation, 19, 25
artery, 18, 19, 20, 21, 25, 31, 36, 38, 43, 44, 45, 52, 59, 76, 78
assessment, vii, viii, 1, 2, 29, 42, 46, 47, 48, 80, 99, 149
astrocytoma, 56, 57, 69

asymptomatic, 51, 57
atherosclerosis, 52
atrophy, 6, 23, 46, 50, 51, 61, 98
auditory nerve, ix, 73, 105
autonomic nervous system, 118, 119
axons, 6, 13, 16, 17, 42, 43, 44, 46, 74, 82, 86, 93

B

benefits, 65, 102, 116
benign, 6, 56, 57
benign tumors, 57
bilateral, 8, 20, 21, 23, 47, 54, 56, 61, 75, 88, 92, 95, 109, 144, 150
biofeedback, 97, 115
biopsy, 52, 59, 63, 92
blindness, 26, 57, 65
blood, viii, x, 19, 41, 45, 60, 100, 118, 121, 127, 136, 138, 143, 152, 155
blood pressure, 121, 144, 152
blood supply, viii, 19, 41, 45, 60
blood vessels, 136
blood-brain barrier, 127
body composition, 153

body mass index, 147
body weight, 132, 137, 146, 147, 148, 149, 150
bone, ix, 3, 6, 10, 11, 12, 13, 16, 21, 44, 58, 63, 64, 71, 73, 74, 75, 76, 77, 82, 83, 85, 86, 103
bone marrow, 71
brain, x, 12, 20, 21, 42, 44, 47, 52, 88, 92, 107, 117, 118, 123, 124, 125, 126, 128, 131, 134, 136, 137, 138, 139, 140, 142, 143, 144, 145, 146, 148, 151
brainstem, 74, 82, 145, 148
breathing, 123, 124

C

central nervous system, 6, 26, 65, 72, 110, 127, 130, 142, 153
central retinal artery occlusion, 48
central retinal vein occlusion, 48
cerebellopontine angle tumor, ix, 74
cerebral arteries, 20, 44
cerebral cortex, 65
children, 8, 36, 56, 84, 87, 88, 97, 104, 107, 109
classification, 6, 21, 22, 24, 33
cleft palate, 6, 30, 35
clinical application, 72
clinical diagnosis, 87, 91
clinical examination, 29, 45
clinical presentation, 88, 90
clinical symptoms, 110, 124
clinical syndrome, 84
cognitive dysfunction, 39
cognitive impairment, 79
cognitive process, 79
compression, ix, 19, 21, 23, 55, 57, 60, 61, 74, 75, 84, 89
computed tomography, viii, 42
conduction, 80, 81, 83, 90, 92
congenital heart disease, 15

control group, 131, 133, 137, 141
corpus callosum, 7, 16, 51
correlation, 26, 69, 100, 110, 137
cortex, 13, 14, 16, 18, 20, 21, 23, 28, 71, 74, 90, 128
corticosteroids, 54, 60, 88, 90, 91
cranial nerve, vii, viii, x, 1, 2, 31, 41, 42, 73, 75, 90, 92, 109, 117, 118, 119, 124, 144, 145, 153
cyclophosphamide, 54, 92
cyclosporine, 52, 92
cyst, 16, 33, 37

D

defects, 7, 8, 12, 13, 16, 27, 111, 127
deficiency, 7, 16, 92, 107, 123, 154
degenerate, 92, 93, 133
demyelinating disease, 81, 84
demyelination, 54, 60, 87
depression, 11, 26, 39, 60, 123, 125, 129, 130, 137, 145, 147, 152
depressive symptoms, 79
depth, 14, 21, 22, 26, 37
diabetes, x, 52, 92, 118, 125, 143, 144, 148, 150
diabetic patients, 143
diet, 125, 130, 132, 143, 147, 148, 150, 153, 154
differential diagnosis, 19, 87
diffusion, 6, 47
diffusion-weighted imaging, 47
digestion, 121, 125, 129, 131, 132
disability, 39, 65, 79, 80, 99, 116
disease progression, ix, 50, 74
diseases, viii, x, 1, 21, 82, 91, 92, 107, 118, 125, 127, 128, 137, 144
distribution, 84, 87, 147
drainage, 20, 31, 51
drug-resistant epilepsy, 149
drugs, 53, 91, 126, 127, 132, 143

dura mater, 9, 44, 120

E

edema, 23, 46, 50, 53, 57, 58, 75, 83, 84, 91
electrodes, 80, 97, 130, 143
electromyography, ix, 74, 81
endothelial cells, 137
energy, 129, 130, 131, 133, 135, 137, 138, 142, 145, 147, 154
energy expenditure, 135, 145, 154
enlargement, 43, 55, 56, 58, 59
environmental change, 2
environmental influences, 29
epilepsy, 128, 129, 137, 150
epithelium, 3, 5, 9, 14, 22, 23
esophagus, 119, 121, 123, 124
ethylene glycol, 52, 92
etiology, 84, 85, 92, 115, 151
evidence, 56, 57, 58, 82, 84, 85, 87, 89, 105
exposure, 47, 52, 62, 91, 95

F

facial expression, viii, 73, 75, 77, 79, 83, 97, 98, 112
facial muscles, 80, 90, 93, 108
facial nerve, vii, viii, ix, 73, 74, 75, 76, 77, 78, 79, 80, 81, 82, 83, 84, 85, 86, 87, 88, 90, 91, 92, 93, 94, 95, 98, 99, 100, 101, 103, 105, 107, 109, 110, 111, 112, 116
facial palsy, ix, 74, 75, 79, 80, 86, 89, 90, 92, 94, 96, 103, 105, 106, 112, 113, 114
fat, 47, 58, 62, 63, 70, 125, 128, 129, 132, 135, 139, 145, 146, 154
fibers, vii, viii, 9, 16, 18, 22, 42, 46, 47, 48, 57, 73, 74, 76, 78, 81, 94, 119, 120, 121, 129, 132, 153
food intake, 128, 129, 130, 131, 132, 133, 134, 135, 137, 138, 142, 145, 146, 147, 150, 152, 153, 154

foramen, 12, 43, 56, 76, 77, 83, 93
forebrain, 9, 15, 26, 42
formation, 87, 89
fractures, ix, 74, 85, 86, 103
frontal lobe, 11, 17, 21, 63, 64

G

gallbladder, 129, 148
ganglion, 42, 43, 47, 50, 51, 65, 71, 72, 76, 78, 82, 87
gastrointestinal tract, vii, x, 117, 118, 129, 131, 142, 151
glaucoma, 46, 50, 51, 65, 71
glioblastoma multiforme, 57
glioma, 56, 58, 59, 64, 69
glossopharyngeal nerve, 87
glucagon, 135, 144, 145
glucose, 92, 110, 126, 128, 130, 132, 135, 137, 138, 139, 141, 143, 148, 150, 155
glucose tolerance, 132
glycogen, 135, 150
glycosylated hemoglobin, 144
growth, 6, 8, 61, 65, 72, 137
growth factor, 65, 72, 137
growth hormone, 61

H

health, 121, 123, 125
health problems, 123
hemorrhage, 55, 57, 60
herpes, 84, 86, 87, 89, 101, 104, 105
herpes zoster, 84, 86, 87, 89, 104, 105
hippocampus, 17, 30, 137, 140, 147
history, ix, 2, 52, 70, 74, 104, 108, 154
homeostasis, 6, 126, 128, 135, 147, 150
human, 2, 34, 36, 70, 71, 107, 118, 121, 127
human behavior, 36
human body, 118, 127
human immunodeficiency virus, 107

hyperglycemia, 132, 143
hyperinsulinemia, 132, 151
hyperlipidemia, 52, 143
hypertension, 52, 61, 92
hypertriglyceridemia, 132
hypoglossal nerve, 93, 94, 95, 112
hypoplasia, vii, 1, 7, 9, 14, 15, 28, 51, 60, 67
hypothalamus, x, 17, 44, 117, 128, 131, 133, 138, 140, 142, 145, 149, 150
hypothesis, 83, 133, 139

I

imaging modalities, viii, 42
immunoglobulins, 68
immunomodulatory, 54, 55
immunoreactivity, 151
immunosuppression, 89
incidence, 8, 15, 53, 86, 90, 91, 92, 102
infection, 27, 84, 87, 88, 89, 104, 107, 108
infectious disorders, ix, 74
infectious mononucleosis, 92
inflammation, 47, 54, 65, 84, 87, 123
inhibition, 128, 133, 135, 139, 141, 144, 145
injury, iv, ix, 3, 22, 44, 55, 65, 74, 76, 77, 82, 83, 85, 86, 93, 94, 103, 111, 146
insulin, 125, 130, 132, 135, 137, 138, 140, 143, 145, 146, 150, 152, 153, 154
insulin resistance, 132
insulin sensitivity, 130, 132, 137, 138, 143, 150, 154
intracerebral hemorrhage, 25
intracranial pressure, 11, 23
intraocular, viii, 41, 42, 43, 50, 51, 55
intraocular pressure, 51, 55
irritable bowel syndrome, 123
ischemia, 20, 39, 52, 91, 92, 93, 146

L

larynx, vii, x, 117, 119, 121, 122, 123, 124
latissimus dorsi, 94, 113
lead, 8, 20, 23, 25, 50, 55, 57, 60, 96, 125, 136, 150
leptin, 134, 138, 140, 141
lesions, ix, 2, 53, 54, 56, 57, 58, 59, 60, 61, 63, 74, 75, 76, 82, 84, 87, 92, 123, 131, 153
light, 44, 45, 47, 48, 50, 81
liver, vii, x, 117, 121, 126, 128, 129, 133, 135, 140, 150, 154
lumbar puncture, 61, 92
luteinizing hormone, 15
lymphadenopathy, 106
lymphoma, 59, 89

M

macular degeneration, 48, 65
magnetic resonance imaging, viii, 28, 36, 38, 42, 46, 67
major depression, 26, 34, 128, 147
management, 38, 52, 53, 54, 60, 62, 64, 85, 103, 110, 111, 131
mass, 12, 23, 53, 107, 125, 129, 145, 146
medical, ix, 62, 74, 123, 126, 149
medical history, ix, 74
medical science, 126
medication, 51, 52, 61, 85
medicine, viii, 2, 3, 36, 101
medulla oblongata, 124
mellitus, 52, 143, 144, 152, 155
meninges, 12, 42, 88, 92
meningioma, 47, 59, 64
mesenchymal stem cells, 6, 65, 71, 72
metabolic diseases, x, 118, 127, 144
metabolic pathways, 135, 136

metabolism, x, 117, 119, 126, 130, 131, 133, 135, 137, 138, 141, 143, 148, 150, 153
mice, 132, 137, 138, 146, 148, 149, 150, 154
motor fiber, 74, 119, 121
mucosa, ix, 5, 7, 22, 73, 75, 134
multiple sclerosis, 26, 37, 39, 53, 60, 67, 90, 108, 129
muscles, viii, 43, 73, 75, 76, 77, 78, 83, 93, 94, 97, 118, 119, 121, 123, 124, 131, 134
musculoskeletal, 15, 52, 88
musculoskeletal system, 15, 88

N

nerve, viii, ix, x, 2, 3, 9, 21, 27, 29, 41, 42, 43, 44, 45, 46, 47, 48, 50, 51, 53, 54, 55, 56, 57, 58, 59, 60, 61, 62, 63, 64, 65, 66, 69, 73, 74, 75, 76, 77, 78, 80, 81, 82, 83, 84, 85, 86, 87, 89, 90, 91, 92, 93, 94, 95, 98, 99, 100, 101, 109, 111, 112, 113, 117, 118, 119, 120, 121, 122, 123, 124, 125, 126, 127, 128, 129, 130, 131, 132, 135, 136, 137, 143, 145, 146, 147, 148, 149, 150, 152, 153, 155
nerve fibers, 9, 43, 51, 60, 66, 75, 83, 87, 135, 143
nervous system, vii, 1, 3, 105, 110, 119, 120
neural network, 120, 147
neuroimaging, 46, 51, 61, 67, 109
neurological disease, 72, 129
neuronal density, 131, 133
neurons, ix, 6, 9, 13, 65, 73, 83, 89, 127, 130, 131, 133, 137, 138, 140, 141, 142, 145, 147, 149, 151, 153, 154
neuropathy, 48, 50, 51, 52, 54, 55, 58, 62, 67, 68, 87, 91, 106, 107
neuropraxia, 82, 83
neuroprotection, 71, 72
neurosarcoidosis, 91

neuroscience, 33, 99
neurosurgery, vii, 1, 31, 57
neurotransmitters, 125, 135
nuclei, 44, 47, 58, 74, 131, 133, 138, 140
nucleus, ix, 16, 44, 47, 73, 74, 78, 94, 124, 130, 131, 133, 138, 139, 140, 141, 142, 145, 147, 148, 151
nucleus tractus solitarius, 145
nystagmus, 51, 60, 86

O

obesity, x, 61, 118, 123, 124, 125, 126, 127, 128, 129, 130, 131, 132, 133, 135, 137, 138, 139, 140, 141, 143, 144, 145, 146, 147, 148, 149, 150, 151, 152, 153, 154
olfaction, 2, 19, 22, 23, 28, 29, 31
olfactory bulb, vii, 1, 2, 3, 6, 7, 9, 10, 11, 13, 14, 15, 16, 17, 18, 19, 20, 21, 23, 25, 28, 29, 30, 31, 32, 34, 35, 36, 37, 39
olfactory nerve, vii, 1, 2, 3, 9, 18, 21, 23, 26, 27, 28, 29, 30, 32, 38
olfactory tract, 2, 9, 14, 16, 17, 19, 21, 23, 29
oligodendrocytes, 42, 43, 59
ophthalmoscopy, viii, 42, 46
optic chiasm, viii, 41, 44, 47, 48, 54, 58, 59, 60
optic nerve, vii, viii, 23, 41, 42, 43, 44, 45, 46, 47, 48, 49, 50, 51, 52, 53, 54, 55, 56, 57, 58, 59, 61, 62, 63, 64, 65, 66, 67, 68, 69, 70, 71, 72
optic neuritis, 47, 53, 54, 67
orbit, viii, 12, 42, 43, 46, 52, 55, 58, 59, 62, 63
organs, x, 117, 118, 119, 120, 121, 122, 127, 131, 136, 140, 144, 153

P

palate, 3, 7, 29, 76, 119, 122, 123, 124

pancreas, vii, x, 117, 118, 131, 132, 135, 140, 143
paralysis, ix, 74, 79, 80, 83, 84, 85, 86, 87, 88, 90, 92, 93, 94, 95, 96, 97, 98, 99, 101, 103, 104, 105, 109, 110, 111, 112, 113, 114, 115, 116, 123
parasympathetic nervous system, 124, 127
parotid, ix, 74, 77, 85, 100, 107
partial seizure, 151
pathogenesis, 51, 75, 89, 91, 132
pathology, ix, 34, 37, 46, 47, 48, 50, 56, 57, 60, 61, 62, 63, 64, 74, 83
pathway, vii, viii, 17, 41, 45, 46, 47, 50, 56, 57, 58, 65, 68, 69, 81, 121, 133, 139, 140, 141, 145, 153, 154
peptic ulcer disease, 132
peptide, 131, 140, 141, 142, 145, 148, 149, 154
pharynx, vii, x, 117, 119, 121, 122, 124
plasticity, 15, 20, 29, 65, 98, 147
positron emission tomography, 143
prednisone, 88, 101, 105
prognosis, 53, 85, 86, 109
proliferation, 6, 57, 154
proptosis, 55, 56, 57, 58, 64

Q

quality of life, 65, 80, 86, 97, 98, 119

R

radiation, 52, 55, 61, 67
radiation therapy, 67
radiotherapy, 57, 58, 59, 69
reactions, x, 82, 118, 136
reconstruction, 64, 111, 112, 113
recovery, 47, 48, 65, 77, 85, 86, 88, 92, 94, 98, 103
reflex sympathetic dystrophy, 93
regeneration, 6, 65, 71, 72, 83, 93, 111

repair, 30, 66, 86, 93
researchers, 128, 131
resistance, 21, 153
resolution, ix, 28, 47, 71, 74, 85, 100, 103
response, ix, 17, 48, 57, 74, 80, 84, 89, 110
retina, 42, 43, 46, 47, 48, 58
retinal detachment, 47
risk, 21, 22, 37, 52, 53, 55, 67, 76, 79, 83, 93, 94, 101, 127

S

salivary glands, viii, 73
sarcoidosis, 54, 68, 92, 110, 111
schizophrenia, 26, 37, 38, 145
secretion, 131, 135, 143, 144, 145
sensation, ix, 73, 76, 78, 87, 94, 121
senses, 2, 28, 120
sensitivity, 20, 34, 49, 54, 83, 125, 132, 137, 143
signals, x, 82, 117, 125, 126, 128, 131, 133, 134, 135, 136, 138, 140, 141, 142, 144, 154
signs, 55, 57, 58, 79, 81, 85, 89, 90
sinuses, 6, 21, 54, 76
skeletal muscle, 81, 143
skin, 6, 64, 77, 88, 97, 120, 122
stenosis, 7, 28, 30, 32, 37, 96
steroids, 52, 53, 54, 55, 85, 86, 94, 102
stimulation, ix, x, 74, 80, 90, 98, 117, 118, 124, 125, 127, 128, 129, 130, 131, 133, 134, 135, 137, 138, 139, 141, 143, 144, 145, 146, 147, 148, 149, 150, 151, 152, 153, 154, 155
stomach, vii, x, 117, 119, 120, 121, 122, 123, 124, 128, 129, 131, 134, 135, 140, 141, 142, 145
symmetry, 80, 96, 97, 115, 116, 123
sympathetic fibers, 76, 120
sympathetic nervous system, 119
sympathetic system, 119, 121, 124

symptoms, ix, 20, 52, 57, 61, 74, 75, 79, 87, 88, 105, 123
synapse, 47, 74, 78
syndrome, 7, 8, 15, 16, 23, 27, 29, 30, 32, 33, 34, 36, 37, 38, 39, 51, 55, 60, 67, 78, 86, 87, 88, 90, 91, 104, 105, 107, 109, 110

T

techniques, vii, 1, 2, 63, 98, 113
temporal arteritis, 52
temporal lobe epilepsy, 25, 26, 30, 31
tendon, 55, 62, 94, 96, 115
tension, 21, 93, 96, 111
territory, 19, 93, 95
therapy, ix, 51, 54, 55, 57, 58, 61, 62, 74, 88, 90, 91, 92, 97, 98, 102, 115, 116, 135, 149
tissue, 6, 8, 23, 55, 89, 94, 97, 114, 127, 129, 140
transmission, 2, 14, 80, 83
transplantation, 72, 113, 114
trauma, ix, 21, 22, 23, 24, 55, 74, 83, 85, 92, 103
traumatic brain injury, 102
treatment, x, 21, 50, 51, 52, 53, 54, 55, 57, 58, 62, 67, 68, 72, 85, 88, 92, 93, 94, 101, 102, 103, 105, 110, 113, 114, 115, 117, 118, 123, 125, 127, 128, 129, 130, 131, 135, 137, 138, 139, 143, 144, 150
trial, 53, 67, 68, 69, 102, 115, 116, 152
tumor, 22, 23, 47, 56, 58, 59, 60, 64
tympanic membrane, 75, 121
type 2 diabetes, 144, 149, 152, 155

V

vagal blockade, x, 118, 152
vagal stimulation, x, 118, 129, 130, 131, 138, 141, 143, 145, 147, 151
vagus nerve, v, vii, x, 87, 117, 118, 119, 120, 121, 122, 123, 124, 125, 126, 127, 128, 129, 130, 131, 132, 133, 136, 137, 138, 140, 141, 143, 144, 146, 147, 148, 149, 150, 151, 152, 153, 154, 155
vessels, 45, 46, 58, 77, 119, 122, 124
vestibular schwannoma, ix, 74
vestibulocochlear nerve, 78, 87
virus infection, 105
viruses, 89
vision, viii, 42, 47, 48, 49, 50, 51, 52, 53, 54, 57, 58, 60, 61, 65, 70
visual acuity, viii, 42, 47, 48, 50, 53, 56
visual field, viii, 42, 43, 49, 50, 51, 60, 61, 96
visual field test, viii, 42, 43, 49, 50
visualization, viii, 42, 45, 63, 64

W

weakness, 75, 79, 84, 90
weight control, 155
weight gain, 123, 128, 154
weight loss, 61, 125, 129, 130, 131, 135, 138, 142, 144, 149, 152

Z

zygoma, 77, 94
zygomatic arch, 77, 94, 96